U0170210

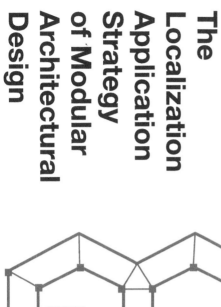

模块化建筑设计的
本土化
应用策略

The
Localization
Application
Strategy
of Modular
Architectural
Design

顾强 著
Gu Qiang

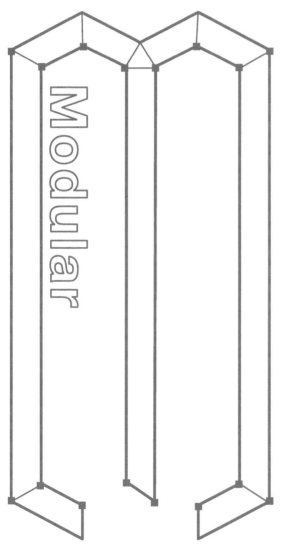

Modular

中国建筑工业出版社

图书在版编目（CIP）数据

模块化建筑设计的本土化应用策略=The
Localization Application Strategy of Modular
Architectural Design / 顾强著. —北京：中国建筑
工业出版社，2022.7（2023.8重印）
ISBN 978-7-112-27551-9

Ⅰ.①模… Ⅱ.①顾… Ⅲ.①建筑设计—研究 Ⅳ.
①TU2

中国版本图书馆CIP数据核字（2022）第109962号

　　本书以装配式住宅进行设计方法及应用规范创新为背景，从模块化建筑的历史发展与现状、模块构成在模块化建筑设计方法、模块化建筑设计和文化传承、模块化建筑设计方法的创新应用、模块化建筑空间的连接方式几个方面进行详细阐述，并结合设计案例，从模块的文化承载方式、模块连接方式、模块在空间设计中的应用方式等角度加以论证，为模块化建筑的个性化发展提供了思路。本书适用于建筑设计、室内设计等相关专业从业者、研究人员等相关人员阅读参考。

责任编辑：张华
书籍设计：锋尚设计
责任校对：王烨

模块化建筑设计的本土化应用策略
The Localization Application Strategy of Modular Architectural Design
顾强　著
Gu　Qiang
＊
中国建筑工业出版社出版、发行（北京海淀三里河路9号）
各地新华书店、建筑书店经销
北京锋尚制版有限公司制版
北京中科印刷有限公司印刷
＊
开本：787毫米×1092毫米　1/16　印张：12¼　字数：230千字
2022年7月第一版　　2023年8月第二次印刷
定价：**68.00**元
ISBN 978-7-112-27551-9
　　（39506）

前言

　　中华文明是中国人民五千多年历史凝聚成的智慧结晶，代表着中国人对自然与世界本源的哲学思考。中国传统文化影响了中国人生活的各个方面，建筑设计自然也包含在内。然而，随着现代技术的进步与设计标准的制定，讲求"人法地，地法天，天法道，道法自然"的中国本土文化如何以当代建筑作为载体，通过适宜的建筑设计方法展现中国本土文化的独有魅力，让中国本土文化元素在当代建筑设计中始终保持活力就是本书作者研究的目的所在。

　　为了实现本土文化元素与当代建筑设计的融合，笔者选取传统模块化建筑作为两者的纽带进行分析研究，并结合中国国情和当代技术手段，探寻模块化建筑设计方法在中国本土化的应用策略。本书所提出的模块化建筑设计方法不同于传统模块化建筑的概念。该设计方法继承了传统模块化建筑的工业化生产特点，但是对于空间模块的使用和文化元素融合方面加以创新，拓宽了模块的设计应用范围，不仅仅限于建筑本身的装配式应用，还涵盖了建筑室内空间的模块置入。因此，本书所使用的模块化建筑设计概念可以重新定义为：一种新兴的建筑结构体系，该体系是以模块空间为单元对建筑及室内空间进行重新建构，根据设计需要在工厂中进行预制生产，完成后运输至现场并组装成为建筑整体或空间模块。

　　模块化建筑设计方法设计出的建筑空间和室内空间能兼具传统模块化建筑所拥有的灵活性、多样性等优点和本土文化的承载性。本书提出的模块化建筑设计方法在不同的空间尺度下，应用模式也不相同：

Preface

The Chinese civilization is the crystallization of the wisdom of Chinese people during more than 5,000 years history. It represents the philosophical thinking of Chinese people on the nature and the origin of the world. Chinese traditional culture has influenced every aspect of Chinese people's life, including architectural design. However, with the development of modern technology and design standards enactment, whether the traditional Chinese culture, emphasizes the thinking of "harmony between man and nature" can still take contemporary design as the carrier, through appropriate building design method to show the unique charm of Chinese traditional culture, keep Chinese traditional cultural elements alive in contemporary design is the purpose of the author.

In order to realize the Chinese localization and cultural inheritance of modular architectural design method, the author selects traditional modular architecture as the link of the two for analysis and research, and combines China's national conditions and contemporary technology means to explore the localized application strategy of modular architectural design method. The modular architecture design method proposed by the author is different from the traditional concept of modular architecture. This design method inherits the industrial production characteristics of traditional modular buildings, but innovates in the use of space modules and the integration of cultural elements, and broadens the application scope of module design. It is not only limited to the assembly application of the building itself, but also covers the module placement of the building interior space. As a result, modular architecture design concept in this book can be redefined as: a new building structure system, the system is based on module space unit to reconstruct of architecture and interior space, according to the requirements of the design for precast production in the factory, and then transported to the site and assembled into the building as a whole or space module.

The architectural space and interior space designed through the modular architectural design method can combine the advantages of traditional modular architecture (flexibility and diversity) as well as the bearing capacity of local culture. The modular building design method proposed in this research has different application modes at different spatial scales:

1. 基于建筑空间尺度：创造体量各异的"盒"空间。模块化建筑设计方法是将建筑主体分为几个建筑模块，在工厂内完成组装和清洗后，将"盒"空间运输到施工现场进行组装。模块化建筑设计方法结合了设计、工厂制造、现场施工和验收等环节。每个模块会根据具体设计方案包含一到多个房间，通过这种方法来提高从建筑设计到建筑施工的整体效率。

2. 基于室内空间尺度：建立独立室内空间的视点和纽带。在建筑空间中插入小的"盒"空间，在区域内创造不同的视觉中心点。以空间与空间的交通流线为纽带进行衔接，以文化造型载体作为独立室内空间的视觉中心点，从而达到牵引和吸引视线的目的。

本土传统文化元素如何与当代建筑设计相融合？解决这个问题的设计方法不仅适用于中国，也适用于其他一些拥有悠久历史的国家。每种本土传统文化都有自己的特点，每种文化都是激发设计师灵感的元素。如果中国传统文化能够被当代建筑设计所包容，那么其他与中国有相似情况的传统文化也可以以本研究为例，作为融合本土文化的建筑设计方法之一。找出本土文化的优势，确定本土文化在当代建筑设计中的定位，将有助于传统文化的继承与发展，同时增加建筑设计方案在当地的可接受程度。

1. Based on the architectural spatial scale: to create "box" spaces with different volumes. Modular building design method divides the main body of the building into several building modules. After finishing and cleaning in the factory, the "box" space is transported to the construction site for assembly. The process of modular building design method includes design, factory manufacturing, site construction and acceptance check. Each module will have one or more rooms depending on the specific design scheme, improving the overall efficiency from architectural design to construction.

2. Based on the interior spatial scale: to establish the viewpoint and link of independent interior space. small "box" spaces are inserted into the building space to create different visual centers within the area. The traffic flow is used as the link to connect different space, and the cultural modeling carrier is used as the visual center of the independent interior space, so as to achieve the purpose of traction and visual attracting.

How to integrate local traditional cultural elements with contemporary architectural design? The design method to solve this problem is not only applicable to China, but also to other countries that have a long history. Each local traditional culture has its own unique characters, and each culture is an element that inspires designers. If Chinese traditional culture can be combined with contemporary architectural design, then some other traditional cultures with similar conditions can also take this research as an example, as one of the architectural design methods integrating local culture. Finding out the advantages of local culture and determine the orientation of local culture in contemporary architectural design will contribute to the inheritance and development of traditional culture, and at the same time increase the acceptability of architectural design schemes in the local area.

目录

Contents

1 模块化建筑理论综述

1.1 模块化建筑概述

模块化建筑是建筑工业化高度发展的结果,其核心概念就是标准化的预制装配式空间模块。它是以一个"房间"为基本的预制构件,在工厂中完成墙体与楼板的连接,制成箱体模块形状的预制整体构件,然后再将这些预制整体构件运输至施工现场,之后如同"搭积木"一样将预制构件与其他现制构件拼装在一起建成建筑体。但是这种"房间模块"不再是一种建筑材料,而是一种空间模块。这是建筑高度工业化的产品,具备良好的完整性。(图1-1)

虽然传统模块化建筑和模块化设计的概念有所差异,但是传统模块化建筑的历史与发展过程为模块化建筑设计方法提供了理论与技术基础。因此,传统模块化建筑的理论综述梳理对于模块化建筑设计方法的应用策略研究具有必要性。

图1-1 某预制装配建筑构件
Figure 1-1 A Prefabricated Building Component Example

1 Literature Review of Modular Building

1.1 Overview of Modular Architecture

Modular building is the result of the high development of building industrialization. Its core concept is the standardized prefabricated space module. The "room" is the basic element of prefabrication. First, in the factory the wall and floor of the room are assembled together to make the prefabricated box, then bring the whole components of the box as well as "building blocks" together to the construction site, for further assimblation. This "box room" is no longer just a building material, but a space module. It is a high-end product of construction industrialization, with its own high integrity. (Figure 1-1) Although the concepts of traditional modular architecture and modular design are different, the history and development of traditional modular architecture provide theoretical and technical basis for modular architecture design methods. Therefore, it is necessary to review the theory of traditional modular architecture for the research on the application strategy of modular architecture design method.

1.2 模块化建筑的发展

　　模块化集成建筑有着相当悠久的
历史。模块化建筑的发展可以被认为
是把建筑建造从"修建"的模式转
入"制造"的模式。修建与制造的区
别何在？修建是一个设计与建造分离
的过程；而制造则是将设计与建造融
合在工业生产环节，即生产标准化预
制配件，在现场进行组装，精准而严
格。早期的模块化建筑先行者希望建
筑业能够如汽车行业一样进行工业
化制造，进而发展出模块化建筑的
雏形。

　　纵观建筑工业化的发展历史，特
别是工业化住宅的发展，重要的契机
和推动力主要来自以下几个方面：

图1-2　工业革命时期，大批农民涌向城市
Figure 1-2 During the Industrial Revolution, Farmers
Flocked to the Cities in Great Numbers

1.2.1 工业革命

　　工业革命推动了相关技术的进
步，现代建筑材料和制造技术得到了
长足发展。与此同时，城市发展也使
大批农民聚集到城市中，推动城市规
模急速扩张，继而导致城市住房供应
严重不足，所以工业革命的到来，让
当时的城市亟须快速补充住宅缺口。
模块化建筑的快速高效性能够满足当
时社会发展的需要，因而工业革命成
了模块化建筑发展的诱因。（图1-2、
图1-3）

1.2 The Development of Modular Buildings

Modular integrated buildings have a long history. The development history of modular architecture can be seen as the shift from "build" mode to "manufacture" mode. What's the difference between building and manufacturing? Building is a process of separating design from construction. Manufacturing is like producing standardized modular parts in a factory, on-site assembly, precise and strict. The early modernist forerunners hoped that the construction industry could be industrialized like the automobile industry, so they developed the prototype of modular architecture.

Throughout the development history of building industrialization, especially the development of industrial housing, the important opportunities and driving forces mainly come from the following aspects.

1.2.1 Industrial Revolution

Technological progress has led to the development of modern building materials and techniques. At the same time, urban development also led to the concentration of a large number of farmers to cities, which promoted the rapid development of urbanization, resulting in a serious shortage of urban housing supply and heavy demand for housing. Therefore, the arrival of the industrial revolution made it urgent for cities at that time to quickly fill the housing gap. The rapid and high efficiency of modular buildings could meet the needs of social development at that time, so the industrial revolution became the inducement for the development of modular buildings. (Figure 1-2, Figure 1-3)

图1-3　工业革命的重要成果：第一座装配式大型公建——伦敦水晶宫
Figure 1-3 The First Large Prefabricated Public Building — the Crystal Palace in London

1.2.2 战争与灾难引发的需求

第二次世界大战后，建筑工业化迎来了真正的高速发展。由于战后的欧洲各国及日本等国面临严重的住房紧缺问题，各国政府迫切需要解决住宅问题，因而促进了模块化建筑的发展。例如，法国的现代建筑大师勒·柯布西耶便曾经构想房子也能够像汽车底盘一样工业化成批生产。他的著作《走向新建筑》奠定了工业化住宅、居住机器等最前沿建筑理论的基础。在此期间，为促进国际的建筑产品交流合作，建筑标准化工作也得到很大发展。

乌兹别克斯坦共和国首都塔什干就是因大灾难导致城市重建进而大规模使用装配式建筑的典型。1966年的一场大地震让古都塔什干一夜之间三分之一的生活区域被毁，30万人无家可归，城市重建迫在眉睫。苏联政府在之后的两年中，用工业化方式对城市进行了快速重建，修建了2300万平方英尺的住宅和15所学校，其中60%的住宅和70%的学校都是预制装配式建筑，塔什干从而成为工业化重建的城市典型。（图1-4）

图1-4 战后重建背景下勒·柯布西耶基于模块化系统和标准化的住宅单元，1952年
Figure 1-4 Le Corbusier's Residential Units Based on Modular System and Standardization in the Context of Large-scale Post-war Reconstruction, 1952

1.2.2 Demands Arising from Wars and Disasters

The real rapid development of construction industrialization began after the Second World War. European countries, Japan and other countries had serious housing demand, which required an urgent solution to the housing problem and therefore promoted the development of prefabricated buildings. For example, Le Corbusier imagined that houses could be manufactured on an industrial scale, like the chassis of a car. His book *Towards New Architecture* laid the foundation of the most advanced architectural theories such as industrial housing and living machine. During this period, in order to promote the international exchange and cooperation of building products, building standardization has also been greatly developed.

Tashkent, the capital of the Republic of Uzbekistan, is a good example of prefabricated buildings which were rebuilt after a major disaster. In 1966, a massive earthquake destroyed a third of the living area of Tashkent overnight, leaving 300,000 people homeless and severely destroying almost all of the city's cultural relics. Over the next two years, the Soviet government rapidly rebuilt the city on an industrial scale, 23 million square feet of housing and 15 schools were built, 60% of housing and 70% of the schools were prefabricated building. Tashkent became a model city for industrial reconstruction. (Figure 1-4)

1.2.3　共产主义与乌托邦思想主导的城市建设

　　共产主义与乌托邦思想主导的城市建设的代表主要集中在以苏联为典型的东欧国家。在乌托邦思想的主导下，城市建设急速扩张。苏联通过不断将农民转化为工人阶层，快速建立起一个工业文明社会。

　　这一时期苏联的建筑工业化得到了很大的发展。苏联政府早在20世纪30年代就开始为工业建筑推行建筑构件标准化和预制装配方法。第二次世界大战后，因为需要修建大量的住宅、学校和医院等，因此定型设计和预制构件也逐渐发展起来。1958—1962年，两三种定型单元的"经济住宅"开始采用工厂化生产；1963—1971年，适用于不同气候区的定型单元设计种类增加到10种。（图1–5～图1–7）

　　在一个多世纪的发展中，不同的时期伴随不同的技术水平和时代需求，工业化住宅呈现的特征也各不相同。在建筑工业化进程中，标准化模数系统的探索意义重大，标准化程度越高则意味着成本的降低和工期的缩短。标准模数还必须适应当时当地的施工条件，符合装配化和机械化的特点，由此才能灵活地满足互换要求。模块建筑体本身也要具有极强的适用性以满足各种建筑类型的需求。

图1–5　20世纪60年代的莫斯科展会：埋入导管的混凝土预制板，镶嵌水管的单元墙体
Figure 1–5 1960s—Moscow Exhibition: Precast Concrete Panels Embedded in Ducts, Unit walls Inlaid with Water Pipes

图1–6　莫斯科混凝土墙制作传送带：完成制作，震荡和砂浆磨平
Figure 1–6 Moscow Concrete Wall Production Conveyor Belt: Complete Production, Shock and Mortar Grinding

1.2.3 Urban Construction Led by Communism and Utopia

The representatives of the urban construction dominated by communism and Utopia are mainly concentrated in the Eastern European countries with the Soviet Union as the typical one. Under the guidance of Utopian ideas, urban construction expanded rapidly. The Soviet Union rapidly established an industrialized civilization by increasing the number of working class and reducing the number of peasants. During this period, the construction industrialization of the Soviet Union has been greatly developed.

This started with the standardization of building components and prefabricated assembly methods in industrial buildings in the 1930s. After the World War II, stereotyped design and prefabricated components were also developed gradually to build a large number of homes, schools, hospitals, etc. From 1958 to 1962, "economic housing" with two or three types of fixed units began to adopt factory production. From 1963 to 1971, the number of types of design for different climate zones was increased to 10. (Figure 1–5~Figure 1–7)

In more than a century of development, different periods had different levels of technology and the needs of the times, the characteristics of industrial housing are also different. During construction process of industrialization, standardization modular system exploration is of great significance, the higher degree of standardization means lower cost and shorter construction period. The standard module must also adapt to the specific local construction conditions, conform to the characteristics of assembly and mechanization, and be more flexible to meet the requirements of interchange. Module of building itself should also have strong applicability to meet the needs of a variety of building types.

图1-7 当时典型的公寓模块，现场装配方式
Figure 1-7 Typical Apartment Module at That Time, on-site Assembly Method

1.2.4 建筑工业化1.0时代

在建筑工业化1.0时代，模块化建筑的主要特点为：快速、大量、廉价、材料单一。大部分设计和建造都相对比较粗糙，但也出现一些技术与艺术结合的例子。这也是真正意义上把建筑设计与工业化进行融合的开端。比如，著名建筑马赛公寓、蒙特利尔栖息地67号（Habitat 67）和芝加哥马利纳城（Marina city），又称"玉米楼"。

（1）蒙特利尔栖息地67号，1967年

蒙特利尔栖息地67号由354个完全预制的独立住宅"盒子"组成，每个基础单位的尺寸为38英尺×17英尺，厨房卫生间也是提前预制的模块。栖息地67号的创新之处在于综合了郊区花园式住宅与城市高层公寓。建造方式在当时被称为"三维模数建造系统"。这座名为"Habitat 67"（栖息地67号）的钢筋混凝土模块化建筑充分发挥了"盒子"作为一种结构形式和建筑造型手段的作用，创造出了前所未有的建筑形象。（图1-8）

栖息地67号的设计理念源于法国作家安东尼·德·圣埃克苏佩里（Antoine de Saint Exupéry）的著作《人类的大地》（*Terre Des Hommes*）。《人类的大地》教给人类如何才能更好地利用自己栖息的土地来生活。栖息地67号遵循着这一理念，将自然环境与建筑设计相融合，为当地居民带来"天堂"般的体验。栖息地67号是建筑师将建筑设计和工业化制造相结合的伟大尝试。该建筑提供了感性设计与理性机械制造的契合点，为模块化建筑设计的发展提供了新思路。（图1-9）

图1-8 栖息地67号（Habitat 67），1967年，设计师：摩西·萨夫迪（Moshe Safdie）
Figure 1-8 Habitat 67, by Moshe Safdie，1967

图1-9 正在组装的"盒子"
Figure 1-9 The Assembling of the "Box"

1.2.4 The Era of Building Industrialization 1.0

In the era of building industrialization 1.0, the main characteristics of modular architecture were: fast, large amount, cheap, and single material. Most of the design and construction were relatively crude, but there were some examples of combining technology and art. This is also the beginning of the integration of architectural design and industrialization in a real scene. Examples include the famous Apartment Marseille, Habitat 67 in Montreal, and the Marina City in Chicago.

(1) Habitat 67, Montreal, 1967

Habitat 67 Montreal consists of 354 fully prefabricated self-contained "boxes", each base unit measuring 38x17 square feet, with a prefabricated kitchen and bathroom module. The innovation of Habitat 67 was the combination of a comprehensive suburban garden house and an urban high-rise apartment. The build method was called "three-dimensional modular construction system" at that time. Named Habitat 67, the reinforced concrete modular building, uses the box as a structural form and ways of architectural modeling to create an unprecedented architectural image. (Figure 1-8)

The design of Habitat 67 was inspired by the French author Antoine de Saint Exupery's *Terre des Hommes*, also known as the *Le Petit Prince*. *Terre des Hommes* explores how humans can make better use of the land they inhabit. Habitat 67 follows this philosophy by combining the natural environment with the architectural design itself as much as possible to bring a sense of "paradise" to people who live here. Habitat 67 is an attempt by architects to combine architectural design with industrial manufacturing. The completion of the Habitat 67 provides a meeting point between perceptual design and rational mechanical manufacturing, and provides a new idea for the development for modular architectural design. (Figure 1-9)

（2）芝加哥马利纳城，1964年

马利纳城（Marina City）为住宅商业混合建筑，其建筑包含两栋61层、高179米（587英尺）的住宅大楼，因其外形酷似玉米棒而成为美国芝加哥市的地标。建筑师贝特朗·戈德堡（Bertrand Goldberg）是密斯·凡·德·罗的学生。虽然马利纳城的设计思想受到密斯的影响，但是戈德堡对预制、模数及曲线形式的设想成为这个建筑设计项目的最大亮点。不同于以往的纯工业化工厂作业，马利纳城建筑设计方案把建筑设计与工业化结合在一起，利用模数化手段将建筑设计造型分解成可工业化量产的成品结构，是建筑设计与工业化结合的典型案例。（图1-10、图1-11）

在建筑工业化1.0时代出现的建筑设计与工业化生产相结合的建筑作品是当代模块化建筑设计的雏形。融合感性设计思维模块建筑体的出现，让模块化建筑作为文化载体成为可能。寻找合理的设计方法，通过模块化建筑构造展现人文情怀，就是本书的重点。

图1-10　每层由16片完全一样的"花瓣"围绕中间的核心筒体组成
Figure 1-10 Each Layer Consists of 16 Identical "Petals" Around a Central Core

(2) Chicago Marina City, Also Known as the "Corn Building", 1964

The Marina City is a hybrid residential and commercial building, consisting of two 61-story residential towers with 179 meters (587 feet) high, which are Chicago's landmarks due to their distinctive appearance, which resembles the corn. The architect Bertrand Goldberg was a student of Mies van der Rohe. Although the design of the Marina City was influenced by Mies, Goldberg's vision of prefabricated, modular and curvilinear forms is the most important aspect of the project. Different from the pure industrial operations in the past, the architectural design scheme of the Marina City combined the architectural design with industrialization, and decomposed the architectural design model into the finished product structure that can be industrialized and mass-produced through modular methods, which is a typical case of combining architectural design with industrialization. (Figure 1-10, Figure 1-11)

The architectural works combining architectural design and industrial production that appeared in the era of architectural industrialization 1.0 are the rudiments of contemporary modular architectural design. The appearance of modular building with perceptual design thinking makes it possible for modular building to act as a cultural carrier. The focus of this book is to find a reasonable design method and show humanistic feelings through modular building construction.

图1-11 弧形的预制构件
Figure 1-11 Curved Prefabricated Components

1.3 模块化建筑的优缺点

选取模块化建筑设计方法作为研究目标，是经过横向考量其优缺点而做出的选择。结合中国亟待解决的乡村改造国情和政策支持导向，模块化建筑能够更加快速有效地解决中国的乡村改造问题并兼具文化传承的作用。

1.3.1 模块化建筑的优点

（1）施工速度快，相比大板建筑可缩短施工周期50%-70%

（2）工业化程度高（装配程度可达85%以上），修建的大部分工作，包括水、暖、电、卫等设施安装和房屋装修都在工厂完成，施工现场只进行构件吊装、节点处理环节，接通管线就能使用。

（3）混凝土盒子构件是一种空间薄壁结构，自重较轻，与砖混建筑相比，可减轻结构自重一半以上。

（4）对环境污染极少。

（5）方便后期扩建。

1.3.2 模块化建筑的缺点

（1）模块化构件的预制工厂投资大。

（2）运输、安装需要大型设备。

（3）造价较高。

（4）单模块面积受限，建筑适用范围低。

虽然建造盒子构件的预制工厂前期投资大，运输、安装需要大型设备，建筑的单方面造价也较贵（与大板建筑差不多），但是依托地方财政补助和清洁能源的自给自足，模块化建筑的前期投入与后期能源回收费用会降低经济压力，弱化模块化建筑的缺点。

1.3 Advantages and Disadvantages of Modular Architecture

In this research, modular architectural design method is chosen as the research target, which is drawn after considering its advantages and disadvantages. After combining China's urgent national conditions of rural reconstruction and policy support orientation, it is obvious that the characteristics of modular buildings can solve the problem of rural reconstruction in China more quickly and effectively and play a role in cultural inheritance.

1.3.1 The Advantages of Modular Architecture

(1) The construction speed is fast, which can shorten the construction period by 50%~70% compared with large slab construction.

(2) With a high degree of assembly (above 85%), most of the construction work, including installation of water, heating, electricity, sanitation and other facilities and house decoration, is moved to the factory for completion. The construction site only has the remaining components to be hoisted and disposed of nodes, which can be used after connecting the pipeline.

(3) The concrete box component is a kind of space thin-walled structure with light deadweight. Compared with the brick-concrete building, the deadweight of the structure can be reduced by more than a half.

(4) Less pollution to the environment.

(5) Convenient for later expansion.

1.3.2 The Disadvantages of Modular Architecture

(1) Large investment in the prefabrication factory of modular components.

(2) Transportation and installation require large equipment.

(3) High build cost.

(4) The area of single module is limited, and the application range of building is low.

The up-front investment in a prefabrication factory to build the box components is too expensive. Large equipment required transportation and installment, and the building cost is expensive (similar to large slab construction), as well. However, relying on local financial subsidies and the self-sufficiency of clean energy, the early investment and late energy recovery cost of modular building will reduce the economic pressure and weaken the shortcomings of modular building.

1.4 模块化建筑的主要构件类型与组装方式

1.4.1 模块化建筑的主要构件类型

模块化建筑构件类型可分为：有骨架盒子构件体系和无骨架盒子构件体系。有骨架盒子构件体系通常用钢、铝、木材、钢筋混凝土制作骨架，以轻型板材围合形成盒子。无骨架盒子构件体系一般用钢筋混凝土制作，每个盒子分别由6块平板拼成，其目前最常采用的是整浇成型的办法，这会使它保持极高的刚度。

（1）有骨架盒子构件体系

有骨架盒子构件体系主要包括空体框架、有平台框架、筒体结构等，较常用钢、铝、木材、钢筋混凝土作为骨架，用轻型板材围合形成盒子，包括：钢龙骨、顶板、底板、外墙板、内墙板、室内装饰等部分。（图1-12、图1-13）

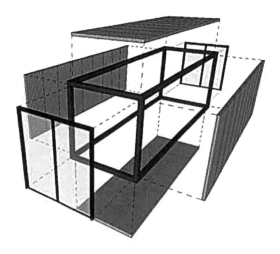

图1-12　有骨骨架体系组成示意
Figure 1-12 Schematic Diagram of Skeleton System Composition

1.4 Main Component Types and Assembly Methods of Modular Building

1.4.1 Main Component Types of Modular Buildings

Modular building component types can be divided into: frame box component system and frame less box component system. The skeleton of frame box component system is usually made of steel, aluminum, wood, reinforced concrete, and enclosed with light plate to form the box. The frame-less box component system is generally made of reinforced concrete, and each box can be made of six flat plates. The most common method is the whole casting molding, which allows its stiffness to maintain a high strength.

(1) The Frame Box Component System

The frame box component system mainly includes empty frame, platform frame and cylinder structure. It usually uses steel, aluminum, wood, reinforced concrete as a skeleton, with light plate enclosed to form a box, including: steel keel, roof, floor, outer wall panel, inner wall panel, interior decoration and other parts. (Figure 1-12, Figure 1-13)

图1-13 骨架体系组成
示意：幕墙工程安装
Figure 1-13 Frame System
Composition Schematic-curtain Wall Engineering Installation

（2）无骨架盒子构件体系

无骨架盒子构件体系通常适用于低层、多层和小于等于18层的高层建筑，一般采用混凝土浇筑而成。这是目前最常采用的整体浇成型方法，可以极大地增强其刚度。（图1－14、图1－15）

两种体系相比较而言，有骨架盒子构件体系因有骨架支撑，墙壁的材料可以选择比混凝土更轻便的材料，所以模块的自重更轻，运输与建造成本更加低廉。此外，有骨架盒子构件体系相较而言可以制造的盒子空间更大，可以根据室内和建筑等不同空间尺度进行置入，灵活性更高。本书探讨的模块化建筑设计方法也是以有骨架盒子构件体系为基础进行设计方法的探讨。

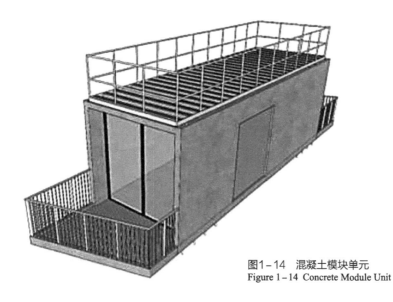

图1－14　混凝土模块单元
Figure 1－14 Concrete Module Unit

(2) The Frame Less Box Component System

The frame less box component system is usually suitable for lower storey, multi-storey and less than or equal to 18-storey high-rise buildings, which are generally made of concrete pouring. At present, this method is the most commonly used whole casting molding method. It can greatly enhance its stiffness. (Figure 1-14)

Compared with the two systems, the frame box component system can choose lighter wall materials because of the scaffold support, and thus the module has lighter weight and lower transportation and construction cost. In addition, the framework box component system can produce a larger box space, which can be placed according to different spatial scales such as indoor and architectural, with higher flexibility (Figure 1-15). The modular building design method discussed in this book is also based on the frame box component system.

图1-15　无骨架混凝土
模块——中银舱体大楼
Figure 1-15 Frame Less
Concrete Module Nakakin
Capsule Tower

1.4.2　模块化建筑的组装方式与构造

传统模块化建筑的组装方式根据模块构件组装支撑方式的不同，主要分为两大类型：模块自重组装和模块框架组装。随着技术的进步，模块化建筑组装方式也涌现出了新型模块化建筑构造方式。

（1）模块自重组装类型

模块自重组装类型的主要组装方式有上下模块重叠组装、模块构件相互交错叠置两种。此类型对模块的自身属性有限制，低自重和高刚度是符合该种组合方式的首选条件。该类型的优势是更简化的施工程序，工厂配套生产工序、成本都相对低廉。缺点是不适用于建造高层建筑。层数越高，工业化生产成本和建造成本就越高。

日本设计师黑川纪章设计的中银舱体楼是模块自重组装类型的典型范例。黑川纪章在设计楼体时，在建筑的东侧和西侧两个不同方向分别采用了上下模块重叠组装和模块构件相互交错叠置的形式。这座建筑是模块化建筑史上里程碑式的作品，极大地推动了模块化建筑的发展。（图1-16）

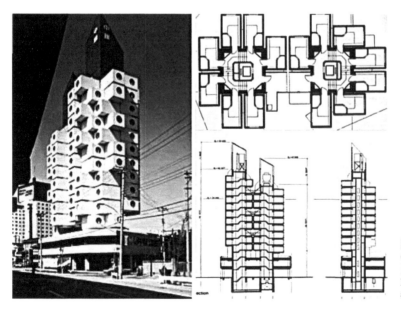

图1-16　Nakagin Capsule Tower 1972 日本中银舱体大楼 Figure 1-16 The Nakagin Capsule Tower, 1972

1.4.2 Assembly and Construction of Modular Buildings

According to the different modes of module components assembly and support, the traditional modular building assembly can be mainly divided into two types: module self-weight assembly and module frame assembly. With the development of technology, there also emerged a new modular building construction method.

(1) module self-weight assembly types

The main types of module self-weight assembly are the overlapping assembly of upper and lower modules and the interleaving and overlapping of module components. These types of module assembly have limitations on their own attributes, and low dead weight and high stiffness are the first choice for this combination. The advantage of this type is that the construction procedure is relatively simple, the factory supporting production process and the cost is relatively low. The disadvantage is that it is not suitable for the construction of highrise buildings. The more the number of floors, the higher cost of the industrial production and construction.

The Nakagin Capsule Tower designed by the Japanese designer Kurokawa was a typical example of module self-weight assembly type. When Kurokawa designed the building body, he adopted the form of overlapping assembly of upper and lower modules and overlapping of module components in the east and west directions of the building respectively. This building was a landmark in the history of modular architecture, which greatly promoted the development of modular architecture. (Figure 1-16)

（2）模块框架组装类型

模块框架组装类型的主要组装方式依据框架种类的不同而有所区分。主要分为预制板材组装方式、框架结构组装方式，以及筒体结构组装方式。此类型对模块的自身属性要求不高。该类型的优势是建设灵活性高，能够根据建筑种类和环境复杂情况自由调节、组装。缺点是施工程序相对更复杂，工厂配套生产工序不能一成不变，导致生产成本变高。

图1-17为模块框架组装方式的实景构造图。通过图片可以看出，建筑模块的支撑力主要来自四周的框架，类似于混凝土建筑的房梁与柱子。建筑模块框架组装的本质就是通过框架外力支撑起模块本身，并按照建筑师的设计架构组合成一个完整的建筑体。

图1-17 框架结构支撑模块构造图
Figure 1-17 Structure Diagram of Frame Structure Support Module

(2) The module frame structure assembly type

The main modes of module frame structure assembly types are distinguished according to different types of frame. It is mainly divided into prefabricated plate assembly mode, frame structure assembly mode and cylinder structure assembly mode. This type does not require much of the module's own attributes. The advantages of this type are high construction flexibility, which means the assembly of the module can be adjusted freely according to the type of building and the complexity of the environment. The disadvantage is that the construction procedure is more complex, and the supporting production process of the factory cannot be unchanged, which leads to higher production costs.

The figure 1-17 is the real construction diagram of the module frame assembly mode. As can be seen from the picture, the supporting force of the building modules mainly comes from the surrounding frame, similar to the beams and columns of a concrete building. The essence of building module frame assembly is to support the module itself through the external force of the frame and combine it into a whole building body according to the architect's design framework.

1.4.3 新型模块化建筑组装方式

随着技术的不断革新，各种兼顾轻巧、环保、高刚度等优点的新型建筑材料被研发出来，如秸秆压强板。这些新型建筑材料的出现，解决了低自重和高支撑力的矛盾点，弱化了模块自重组装类型的缺点。这也为新型模块化建筑的组装方式打开了新局面，拓宽了应用领域。

现有的新型模块化建筑组装方式可以说是传统模块自重组装类型的升级版。这种新的组装方式主要以轻型盒子为主，即作为建筑主体的轻型模块，同时承担建筑体支撑的任务。建造过程如同搭积木，每个模块之间利用焊接、铆钉等方式进行衔接即可。施工过程简捷，运输成本低，能够根据设计师的想法灵活搭配。工业化生产固定模块结构，不需要针对每个项目重新制作模具，减少了生产成本。因此，新型模块化建筑组装方式融合了两种传统模块化建筑组装方式的优点，弱化或规避了它们的缺点，使模块化建筑能够更加适应复杂多变的设计思路和地理环境，为模块化建筑设计方法承载本土文化提供了必要的技术条件。（图1-18）

1.4.3 New Modular Building Assembly Mode

With the continuous innovation of technology, a variety of lightweight, environmental friendly, high stiffness and other advantages of new building materials have been developed, such as straw pressure plate. The emergence of these new building materials solves the contradiction between low dead weight and high supporting force, and weakens the shortcomings of module dead weight assembly type. This also opens up a new situation for the new modular building assembly method and widens the application field.

The existing new modular building assembly mode can be recognized to be an upgraded version of the traditional module self-weight assembly type. The main type of this new assembly mode is light box assembly mode, that is, light modules are used as the main body and the supporting structure of the building at the same time. The construction process is like building blocks. Each module is connected by welding, rivet and other ways. The construction process is simple, the transportation cost is low, and can be flexibly matched according to the designer's idea. Industrial production of fixed module structures does not need to re-make the mold for each project, which reduces the production cost. Thus, the new modular architecture assembling approach combines the advantages of two traditional modular building assembly type, and weaken or avoid their shortcomings, and thus, the modular architecture can adapt better to the complex and changeable design ideas and geographical environment. It provides the necessary technical conditions for modular building design method to bear native culture. (Figure 1-18)

图1-18 轻型盒子装配式建筑建造施工图
Figure 1-18 Pictures of Light Box Assembly Building Construction

1.5 模块化建筑在中国

1.5.1 中国模块化建筑的过去与现在

中国在1979年就开始了模块化建筑的探索与尝试，陆续在青岛、南通、北京等地试建了几栋模块化房屋。但是与当时其他国家面临的情况一样，这些建筑更偏向于解决工业化快速发展所带来的城市住房不足的问题，并没有考虑主观建筑设计思维和客观建筑体本身相结合的问题。

随着经济的不断发展，提高乡镇居住水平也被中国定为需要重点解决的民生问题之一。根据中国国情，能够满足快速、高效、环保要求的模块化建筑设计方法承担起了振兴乡村和承载地域特色文化的重任。因此，中国设计师在模块化建筑领域进行了设计尝试，以期摆脱之前只求速度、实用，不求美观、文化内涵的快餐式建筑。许多优秀的多层模块化建筑作品被设计出来，为中国当代模块化建筑领域提供了宝贵的实践经验。比如：张家口官厅公共艺术小镇项目、江苏兆智–IMBox移动度假屋设计项目等。

（1）张家口官厅公共艺术小镇项目

官厅公共艺术小镇的观湖集装箱酒店，以集装箱为建筑构件，高标准打造，含套间及标间，内部配置一应俱全。经过特殊处理的集装箱，具备绝缘隔热、坚固耐用的特点。酒店四面环山、风景优美，非常适合旅游度假。40余栋色彩艳丽的度假屋，犹如一件件精美的艺术品矗立在水库旁边，为水库增色不少。（图1–19）

图1–19 张家口官厅公共艺术小镇项目
Figure 1–19 Zhangjiakou Guanting Public Art Town Project

1.5 Modular Architecture in China

1.5.1 The Past and Present of Modular Architecture in China

As early as 1979, China began to explore and try modular architecture. China has built several modular houses on trial in Qingdao, Nantong, Beijing and other cities. But as the case of other countries, these house were to solve the urban housing shortage problem brought by the rapid development of industrialization, and did not take into account the combination of the subjective architecture design thinking and the objective building itself.

With the continuous development of economy, the improvement of living conditions in towns and villages has also been identified as one of the key livelihood issues to be solved in China. According to China's national conditions, the modular architectural design method, which can meet the requirements of fast, efficient and environmental protection, were adopted to tackle with the important task of revitalizing the countryside and bearing the regional characteristic culture. Therefore, Chinese designers began to try to design in the field of modular architecture, in order to get rid of the previous fast food architecture, which only seeks speed and utility, and does not seek beauty or cultural connotation. Many excellent low-rise modular buildings have been designed, providing valuable practical experience for the contemporary modular building field in China. For example: Zhangjiakou Guanting Public Art Town Project, Jiangsu Zhaozhi-IMBOX Mobile Holiday House Design Project, etc.

(1) Zhangjiakou Official Hall Public Art Town Project

Guanhu Container Hotel, Guanting Public Art Town, is built with containers as building components, with high standards, including suites and standard rooms, complete internal configuration, specially treated containers, with insulation, strong and durable characteristics. The hotel is surrounded by mountains, fresh air and pleasant scenery, which is very suitable for people to travel and spend their holidays. More than 40 colorful holiday houses, like a piece of fine art standing beside the reservoir. (Figure 1-19)

（2）江苏兆智–IMBox移动度假屋设计项目

设计团队仿照大地被自然侵蚀的过程，创造了雕塑性的建筑外观，让天空降临在建筑周围的自然环境之中。建筑主体由模块化结构组成。通过模块化的立面组件和分散体块的共同作用，模糊了建筑的轮廓。木材温暖的质感反射着柔光和金属板在空中交错，使建筑在光线中觉醒。策略的核心部分是将自然融入建筑设计，创造具有私密性的公共花园，唤醒人所有的感官体验。多级露台和空中花园出现在作品的各个层次中，在建筑内部和外部都能感受到自然带来的舒适感。该项目不仅为游客创造了舒适、静谧的住宿环境，同时为植物的生长创造了空间，在立面中交替创造出巴黎式的阳台空间。（图1–20）

图1–20　江苏兆智–IMBox移动度假屋设计项目效果图
Figure 1–20 Jiangsu Zhaozhi– IMBOX Mobile Holiday House Design Project Rendering

(2) Jiangsu Zhaozhi-IMBOX Mobile Holiday House Design Project

The design team created a sculptural facade that mimics the erosion of the earth, allowing the sky to descend on the surrounding natural environment. The main body of the building is composed of a modular structure. Modular facade components and scattered volumes work together to blur the outline of the building. The warm texture of the wood reflects the soft light and the metal plates crisscross the air, awakening the building to the light. The central focus of the strategy is to integrate nature into the architectural design to create a public garden with privacy that evokes all sensory experiences. Multilevel terraces and hanging gardens are present at all levels of the work, providing a sense of natural comfort both inside and outside the building. The project not only creates a comfortable, introverted accommodation experience for visitors, but also creates space for plants to grow, alternately creating Parisian balconies in the facades. (Figure 1-20)

1.5.2　中国模块化建筑的扶持政策

2006年，住房和城乡建设部下发《国家住宅产业化基地试行办法》，要求建筑行业要走出一条科技含量高、经济效益好、资源能耗低、环境污染少、人力资源优势得到充分发挥的新型工业化道路。这份文件为模块化建筑设计方法在中国的推广提供了政策支持。同年，住房和城乡建设部正式颁布了《绿色建筑评价标准》，规范了绿色建筑的评判体系，为模块化建筑设立了行业准则。

2012年5月，财政部发布《关于加快推动中国绿色建筑发展的实施意见》，明确规定对高星绿色建筑给予财政奖励。奖励标准为二星绿色建筑奖励45元/平方米，三星绿色建筑奖励80元/平方米。这些政策的颁布，让模块化建筑既得到政府政策层面的支持还弱化了前期厂房建设带来的资金压力。

2016年，国务院办公厅发布《关于大力发展装配式建筑的指导意见》（以下简称《意见》)，《意见》中明确指出，"力争用10年左右的时间，使装配式建筑占新建建筑面积的比例达到30%。"该《意见》明确了模块化建筑在未来中国建筑业发展的主要地位。

1.5.2 China's Relevant Support Policy for Modular Buildings

In 2006, the Ministry of Housing and Urban-Rural Development of China issued the *Trial Measures for National Housing Industrialization Base*, requiring the construction industry to open up a new industrialization way with high scientific and technological content, good economic benefits, low energy consumptions, less environmental pollution, and full use of human resources advantages. This document provides policy support for the promotion of modular building design methods in China. In the same year, the Ministry of Housing and Urban-Rural Development officially promulgated the *Green Building Evaluation Standard*, standardizing the evaluation system of green buildings and setting up industry guidelines for modular buildings.

In May 2012, the Ministry of Finance of China issued the *Implementation Opinions on Accelerating the Development of Green Buildings in China*, which clearly stipulated that financial incentives should be given to high-star green buildings. The reward standard is 45 yuan per square meter for a two-star green building and 80 yuan per square meter for a three-star green building. With the promulgation of these policies, modular building not only gets the support from the government policy level, but also reduces the financial pressure brought by the early factory construction.

In 2016, the General Office of the State Council of China issued the *Guiding Opinions on Vigorously Developing Prefab Buildings*. The "Opinions" clearly pointed out that " strive to make prefab buildings account for 30% of the newly built floor area in about 10 years". The "Opinions" identified the major role of modular buildings in the future development of Chinese construction industry.

1.6 本章小结

综上所述，模块化建筑设计方法在中国具有良好的发展空间，同时也积攒了一定的实践经验。在乡村改造的大趋势和政府政策导向下，模块化建筑设计利用适应性本土化应用方式，能够突破传统模块化建筑的固有模式，灵活利用空间模块进行设计，提高其在中国市场的认可度。将中国本土文化特色、人文内涵更好地融入模块化建筑设计，会使模块化建筑更好地适应地方，提高可接受程度，成为打开中国市场、提高市场占有率的金钥匙。

1.6 Conclusion

In conclusion, modular architectural design method in China has a good development prospect, with several accumulated practical experience. Under the general trend of rural reconstruction and the guidance of government policies, the market recognition of modular architectural design in China will gradually improve. Modular architectural design can break through the inherent mode of traditional modular architecture by using adaptive and localized application, flexibly use space modules to design, and improve its recognition in the Chinese market. Better integration of Chinese local cultural characteristics and humanistic connotation into modular architectural design will make modular buildings better adapt to local conditions and improve the acceptability, which will become the golden key to open the Chinese market and increase market share.

2 模块化建筑设计法中的模块组成方式

模块化的概念至今没有统一定义，比较统一的理解是模块化是一种设计方法，同时也是一种标准化的方法。在不同领域，模块都具有通用的属性：灵活性和互换性。每个模块都有一个特定的子功能，所有模块按照一定的逻辑组织起来，形成一个系统。

模块化建筑设计法中的模块组成方式继承了模块的基本属性。不同空间模块之间可以通过拼合、插接、分解进行排列组合，同一个模块自身可以分解、变形、移动。模块通过这些方式进行空间上的重组，即模块化建筑设计法。本书选取了一些建筑领域的典型案例，总结并用图解的方式诠释了空间模块的特征与模块之间的组织方式。笔者将在下文进行模块之间的空间组织方式与案例分析。

2.1 模块主要组成方式

2.1.1 分隔

烟台日报社办公空间项目使用的是多个模块分隔、拆解组合成的新空间。该项目将模块变为一个"百变"的空间概念，将空间与功能需求灵活组合。一天中，员工们有多种活动在这个模块中进行：会客、午饭、沙龙交流、设计办公等，如果每一个活动都需要单独的空间，那么就需要较大的空间。根据员工的需求，灵活地分割、拆解、组合模块空间，能在有限的空间内，组织不同的功能。在由模块分解、组合的不同空间模块里工作，员工的专注度和办事效率得到了提高。这是模块化建筑设计方法的典型运用，具有标准化、环保化、经济化等显著特点。

2 Module Composition in Modular Architectural Design Method

There is no unified definition of the concept of modularization, the rather unified understanding is that modularization is a design method, but also a standardized method. Modules have common attributes in different domains: flexibility and interchangeability. Each module has a specific sub–function, and all modules are organized according to certain logic to form an overall system.

The modular architectural design method inherits the basic properties of modules. Different space modules can be assembled, inserted and decomposed for arrangement, and the same module itself can be decomposed, deformed and moved as well. This is the method of the building modules. This book selected some typical cases in the field of architecture, summarized and illustrated the characteristics of spatial modules and the way of organization between each modules. The spatial organization of modules and case analysis are as follows:

2.1 Module Composition Mode

2.1.1 Separation

The office space project of Yantai Daily is separated and disassembled into a new space by multiple modules. The project transform the module into a "fluid" spatial concept, combining spatial and functional requirements flexibly. Throughout the day, employees have a variety of activities in this module: reception, lunch, salon communication, office design, etc. If each activity needs a separate space, it would require larger area. According to the needs of employees, being able to split, disassemble, combine module space flexibly, would allow organizing different functions in a limited space. Working in different space decomposed and combined by modules, the employees' concentration and efficiency have been improved. This is a typical use of modular architectural design method, which has the obvious characteristics of standardization, environmental friendly and economy.

2.1.2 插接

建筑的"插件塔"设计了专门的模块连接构架，不同功能的空间模块可以插在这些构架之中。模块可以拆解的同时又方便组合。可以设想，在一个空间中，需要临时拆除一个6人的空间，这时，可以利用模块易于拆解的特性拆除空间，同时不破坏原有的空间结构。插件塔是一个预制的"插件板"构架系统，可以任意拓展。在结构的框架体系中，可以插入房屋的单元，这些单元可以根据需求放置于不同位置。板材之间用钩锁连接，安装搭接方便。模块化的拼装方式无须大兴土木，省时省料。

2.1.3 移动

"众行顶"是一个有红色顶棚的装置，可以伸缩，折叠时可供市民骑行穿越整座城市；展开时可以覆盖部分街道，连接不同的城市公共空间。众行顶顶部参照常见的模块化拉伸结构建造的大棚，宽度可以容纳10多个人在内部骑行。结构底部装有自行车车轮，方便移动。众行顶展开时，可以延伸至12米，覆盖面积大。众行顶被欧洲等国家用于举办一些临时性的公众活动。模块化可延展、可伸缩、可移动的特性在众行顶建筑中得到了很好的展现。

2.1.2 Connection

The "plug–in tower" is designed with a special modular connection framework in which spatial modules of different functions can be inserted. Modules can be disassembled and assembled easily. It can be imagined that in a space, a space for six people needs to be temporarily dismantled. At this time, the module can be easily disassembled to dismantle this space without damaging the original space structure. The plug–in tower is a prefabricated "plug–in board" architecture system that can be expanded at will. In the frame system of the structure, units of the house can be inserted, and these units can be placed in different locations according to requirements. The plates are connected with hook locks, a convenient installation lap, no need for mass construction, save both time and materials.

2.1.3 Movement

The "Moving Roof" is a retractable red–roofed installation that folds for citizens to ride across the city. When unfolded, it can cover part of the street and connect different urban public spaces. The top of the roof refers to the common modular stretch structure of the greenhouse, which can accommodate more than 10 people to ride on. The bottom of structure is equipped with bicycle wheels for easy movement. When the "Moving Roof" is unfolded, it can extend to 12 meters.The "Moving Roof" is used in Europe and other countries to hold some temporary public events. The characteristics of modular extension, expansion and mobility are well played in the moving roof architecture.

2.1.4 变形

"百变智居"设计过程中，模块可以根据不同功能模式被灵活组织，像抽屉一样被拉拽，打破单一的室内外空间界限。

模块可以根据需要，变换三种不同的模式，家庭、办公、聚会。在家庭模式中，模块的空间划分为客厅空间、卧室空间以及餐厨空间。墙面可以翻折起来，打通卧室空间，满足基本的居住需求。在办公模式中，隔墙可以自由移动变化，将空间分隔为会客空间和办公空间。空间结合桌子，可以形成巨大的工作坊。通过打开墙体与拉出箱体，可以形成聚会空间。客厅中可以找出一个翻折的床板，拉出后可以供多人使用。模块化的设计方式，通过预制工业化的构件和现场装配的特点，可以节省工期，搭建灵活可变的空间。墙面、楼板可翻折，空间可推拉，智居空间的灵活可变性是设计的亮点。

2.1.5 累加

栖息地67号（Habitat 67）是1967年世博会上的展览建筑。世博会之后，这组建筑被保留下来，现在成为加拿大的国家文化遗产。栖息地67号建筑群是将一些预制的住宅单元盒子以参差错落的形式叠加起来，形成一个个复杂的建筑空间。这些盒子错综叠加形成的灰空间，成为立体花园和阳台，既满足了住户的采光需求，又保证了私密性。设计师摩西·萨夫迪（Moshe Safdie）旨在通过这种方式向中低收入人群提供可以复制的、廉价的住房。栖息地67号曾一直被闲置，直到20世纪70年代，加拿大设计师将其中的几间单元盒子改造一新，兼顾了时尚感与舒适性后，越来越多的富人陆续搬进这个社区，栖息地67号也自此变成了高档社区。值得一提的是，栖息地67号的所有居住者共同拥有该建筑的产权。这座建筑群落如今已是宝贵的文化遗产，所以，产权共有对于居住者来说是一件十分骄傲的事情。

栖息地67号模块化堆叠的建筑盒子，在加拿大蒙特利尔圣罗伦斯河畔格外醒目，小山一样的建筑群组极具震撼力。建筑空间模块之间叠加的处理手法值得后辈设计师学习借鉴。这座建筑群是模块化的设计思维在建筑领域具体应用的经典案例。栖息地67号成为当地的地标，也成为设计界的经典案例。

2.1.4 Transformation

During the design process of "transform smart home", modules can be flexibly organized according to different functional modes and pulled like drawers, breaking the boundaries of a single indoor and outdoor space.

Modules can be changed according to the needs of three different modes, home/office/party. In the family model, the module is divided into living room, bedroom and kitchen space.The metope can be fold up, open up the bedroom space, satisfy the use requirement for basic center space. In the office mode, the partition wall can move freely and change. After the partition wall moves, the space is divided into receiving use and office use. The space with table can form a huge workshop space. In party mode, by opening the walls and pulling out the box, a party space can be formed. A folding bed board can be found in the sitting room, which can be used by many people after being pulled out. Modular design, through prefabrication of industrial components and on-site assembly, can build variable spaces and save time. The walls and floors can be folded, the space can be pushed and pulled, and the flexibility of intelligent living space has become the highlight of the design.

2.1.5 Superposition

Habitat 67 was the exhibition building at the 1967 World's Fair. After the Expo, the buildings were preserved and are now a Canadian national heritage site. The Habitat 67 complex is a jumble of prefabricated housing units. These residential units over layed one on top of another, formed complex architectural Spaces. The gray space formed by the overlapping of these boxes becomes three-dimensional gardens and balconies, which satisfied the lighting and privacy needs of residents. Designer Moshe Safdie aimed to provide replicable, affordable housing for low and middle-income people in this way. Habitat 67 staied unused until 1970s, when several Canadian designers refurbished several of its boxes with the feeling of fashion and comfort. After that, more and more wealthy people moved into the neighborhood, and Habitat 67 became upscale. It is worth noting that all the occupants of Habitat 67 jointly own the property rights in the building. The building community has become a valuable cultural heritage, so co-ownership is a matter of great pride for the residents.

Habitat 67's modular stacked building boxes stand out at the banks of the St. Lawrence River in Montreal, Canada. The processing technique of superposition between architectural space modules is worth learning. The building complex is a classic example of the concrete application of modular design thinking in architecture.

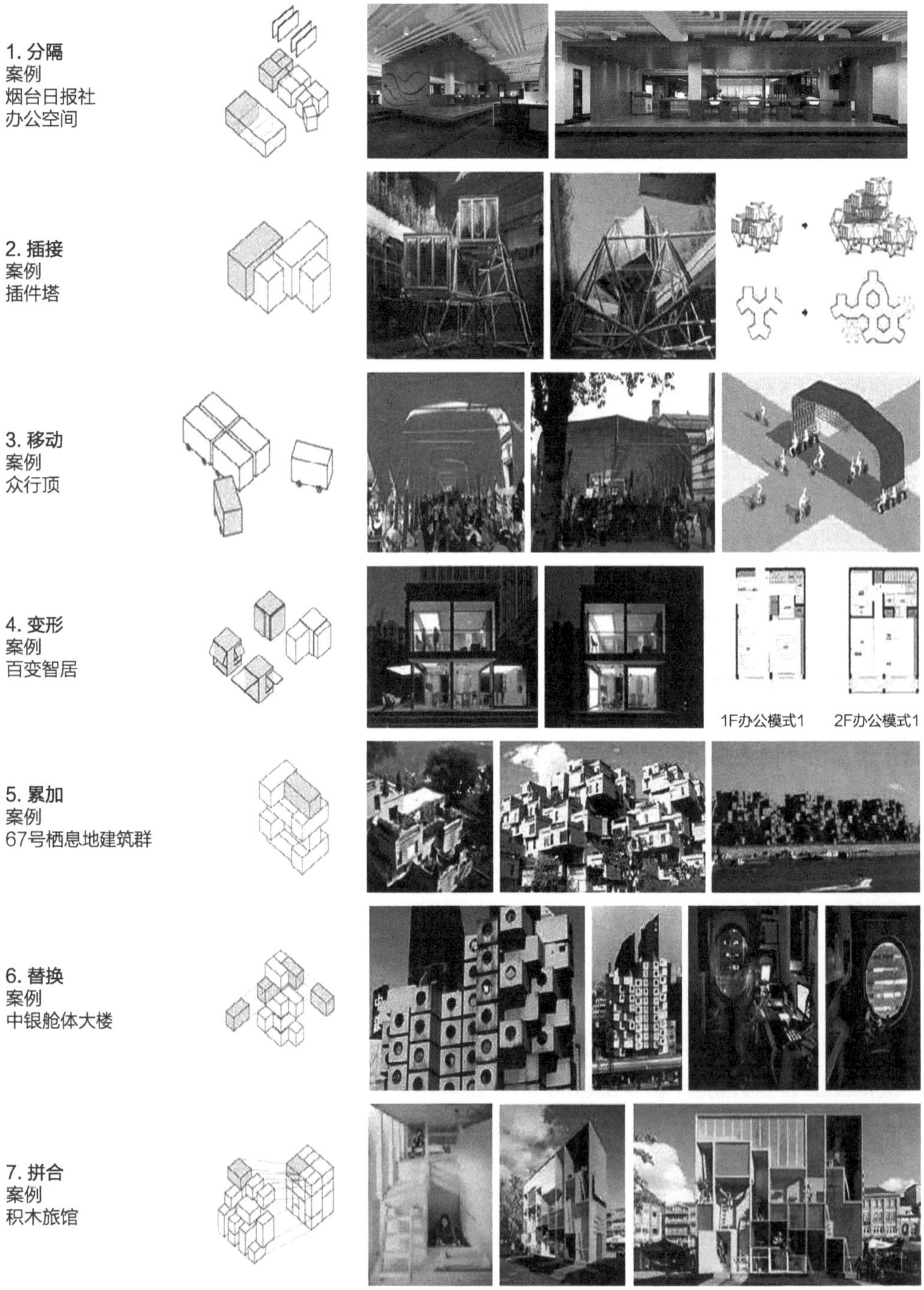

1. 分隔
案例
烟台日报社
办公空间

2. 插接
案例
插件塔

3. 移动
案例
众行顶

4. 变形
案例
百变智居

1F办公模式1　2F办公模式1

5. 累加
案例
67号栖息地建筑群

6. 替换
案例
中银舱体大楼

7. 拼合
案例
积木旅馆

图2-1　模块组成方式及案例示意图

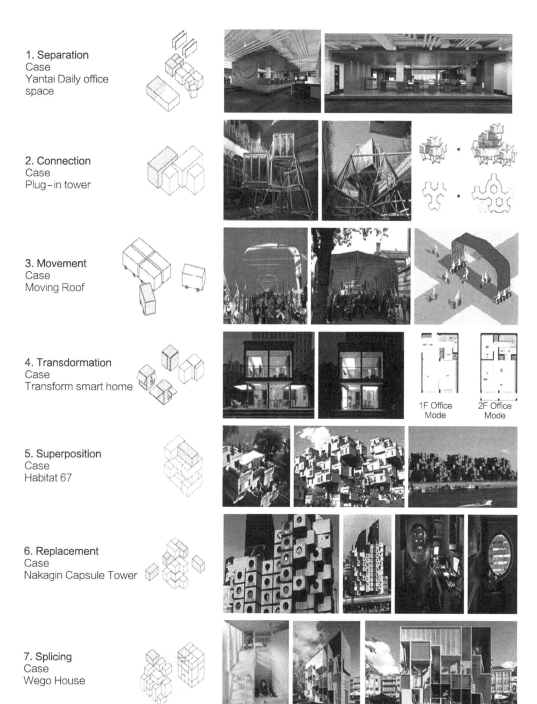

Figure 2-1 Module Organization and Practical Case Diagram

2.1.6 替换

在日本东京繁华的银座区，耸立着一栋怪异的建筑，许多胶囊状的像太空舱一样的建筑盒子，一簇簇摞在一起，科幻未来感十足。这栋建筑就是中银舱体大楼，它曾经象征着日本对未来的幻想。日本新陈代谢派设计师黑川纪章设计了这栋建筑。"新陈代谢派"起源于20世纪60年代，强调万事万物逐渐生长、迭代和衰亡的过程，主张用新技术来解决问题。

2.1.7 拼合

2017年荷兰设计周展出了荷兰建筑事务所MVRDV设计的一款概念居住空间——积木旅馆（Wego House）。整栋建筑看起来就像一组各种颜色拼合起来的积木。不同颜色的积木代表着不同的空间属性。粉色空间中设置了通透的阳台；黄色空间设置了一个可以居住的阁楼；绿色空间设置了可以摆放植物和吊床的场所。不同模块的空间有不同的空间气质，提供不同的功能，比如休憩、居住、展览。一些模块是全部开敞通透的，另一些模块是半透明、半通透的。积木旅馆根据不同人群的不同需求，比如学生、难民、音乐家、画家等，提供不同的休憩、展览、学习等功能。依据这样一个模块化的设计逻辑，每个模块可以根据使用者的不同需求而定制使用。这些模块拼合在一起形成一个居住的建筑体。

2.1.6 Replacement

In the Ginza district of Tokyo, Japan, stands an eerie building, with many capsules shaped like modules stacked on top of each other in a futuristic fashion. The Nakagin Capsule Tower, once symbolized Japan's vision of the future. Japanese metabolic designer Noriaki Kurokawa designed the building. The metabolisms originated around the 1960s, emphasized the gradual growth, iteration and decay of everything, and advocate new technologies to solve problems.

2.1.7 Splicing

During the Dutch Design Week 2017 showcases, Wego House, a concept living space designed by Dutch architecture firm MVRDV is on the show. The whole building looks like a set of colored blocks. Different colored blocks represent different spatial attributes. The pink space is filled with transparent balconies; The yellow space has a habitable loft; The green space provides places for plants and hammocks. Different modules of the space have different spatial temperament, and provide different functions as rest, living, exhibition. Some modules are fully open and transparent; Other modules are semi transparent. Wego House is designed for the needs of different groups of people, such as students, refugees, musicians, painters and so on, to provide different rest, exhibition, learning and other hotel functions. According to such a modular design logic, each module can be customized according to the different needs of users. These modules are pieced together to form a residential building.

2.2 实际应用案例

为了用模块化的思路更好地进行建筑设计，本书作者选取了烟台大学翼之队的设计项目"北方印宅"，作为具体的模块化组成设计实践应用案例。此建筑设计项目参与了2018年中国国际太阳能十项全能竞赛，是将模块化建筑设计法与绿色节能技术相结合的实际案例。

2.2.1 北方印宅项目简介

北方印宅项目使用了模块化建筑、智能家居、立体绿墙、雨水回收系统等技术，采用稻草、竹子等绿色环保建筑材料。在这个项目中，本土化、传统化和现代技术相互碰撞。这种融合的设计旨在给人们营造一种文化体验感和归属感，建造一个新型的、区域性的生态家园。

建筑北面有两层，南面有一层，形成了一个倾斜的屋顶，最大限度地扩大了太阳能电池板所需的面积；建筑的功能形式是一个简单的"L"形，使建筑形体系数和热量消耗最小化。中庭设置自然景观，打开东侧主体，能够呼应外部景观。（图2-2）

北方印宅追求的是一种结合当地传统建筑和现代技术的建筑模式，基于环境和地理的特殊性，探索新的可能性，并将其与传统建筑方法相结合。"利用自然，避免自然的危害"，设计师们希望探索一种新的"本地化生态住宅"。

图2-2　北方印宅入口效果图
Figure 2-2　The entrance rendering of the North Yin House

2.2 Practical Application Cases

For the purpose of using modular ideas for better carrying out architectural design, the author selected the design project "Yin House" of Yantai University Yi-Team, which he participated in, as a specific modular composition design practice case. This architectural design project participated in the 2018 China International Solar Decathlon competition, which is a practical case of combining modular architectural design method with green energy saving technology.

2.2.1 General Introduction of Yin House

Modular building, intelligent home, three-dimensional green wall, watering the vegetable garden with rainwater recycling system, using green materials such as straw and bamboo as raw materials. Here, localization, traditional Ization and modern technology collide with each other. The blend was designed to give people a sense of experience and a sense of belonging to a new, regionally ecological home.

The building has two floors to the north and a layer on the south side to form a single sloping roof that maximizes the area required for solar panels; the building's functional form is a simple "L" shape that minimizes the building body shape coefficient and heat consumption. The atrium sets the natural landscape, and opens the east side body, echoing the external landscape. (Figure 2-2)

It pursues a building model that combines localized, traditional regional architecture with modern technology, based on the particularities of the environment and geography, exploring new possibilities and combining them with tradition. "To take advantage of nature and avoid the harm to nature", the designers hope to explore a new "localized ecological residence".

2.2.2 回归生命建筑，承担文化使命

中华民族的居所是现代桃花源的再现。民居是时间的艺术，空间的灵魂，穿越时间，穿越空间，需要我们不断地探索。古朴典雅，是中国传统建筑的一种独特风格，在中国悠久的历史和文化中有着特殊的意义。它优雅而沉稳，承载着中华民族特有的精神和文化，是中国人民几千年积累下来的一种境界与智慧，可以与现代生态理念相辅相成。

在这个项目的思考中，设计师将传统的建筑环境智慧与现代的生态理念相结合。回归生命的建筑承担着弘扬传统文化的使命。

因此，在庭院布局的基础上，设计师尝试在传统模式的基础上对传统建筑的形态进行创新，并突出其优势，将现代模式和生活模式融入新的建筑形态，创造一种传统与现代结合的新型四合院。整体的单坡屋顶最大化了太阳能电池板所需的面积；公寓简单的"L"形减少了建筑的外形因素和热量消耗。风水强调方位布局和环境模型，整个建筑体营造出一个自然独立的生态系统。建筑部分采用半室外空间和室外环境，合理协调规划。这四个部分，相互穿插，形成一个相对科学、独立的生态有机体。

在室内和室外的装饰中，设计师采用了传统的建筑装饰材料——竹子来制作建筑室内顶棚和部分墙面装饰，从而更加贴合中国传统文化。（图2-3）

图2-3 北方印宅设计项目竹子材料的应用
Figure 2-3 Construction of Bamboo Material of Yin House Design Project

2.2.2 Return to Life's Architecture and Undertake the Mission of Culture

The residence of the Chinese is the reproduction of the modern peach blossom source. Folk dwelling is the art of time, the soul of space, which passing through time, crossing the space, we need to constantly explore. Simple and elegant, it is a unique form of Chinese traditional architecture. It has a special significance in China's long history and culture. It is elegant and calm, bearing the unique spirit and culture of the Chinese. A realm and wisdom is the crystallization of human wisdom for thousands of years, and can complement the modern ecological concept.

In thinking about this project, the designers combined traditional wisdom of the built environment with modern ecological ideas. Returning to the life of architecture is to undertake the mission of culture.

Therefore, based on the layout of the courtyard, the designers tried to innovate the shape advantages of traditional architecture on the basis of the traditional model, and integrated modern patterns and living patterns into the new architectural form to create a new combination of traditional and modern courtyard house. The overall single-sloping roof maximizes the area required for the solar panels; the simple "L" shaped flat reduces the building's form factor and heat consumption. The azimuth layout and environmental model are emphasized in Feng Shui, creating a natural independent ecological circulation system. The building partly adopts a semi-outdoor space and an outdoor environment for reasonable coordination and planning. These four parts are interspersed with each other based on the main body of Feng Shui, forming a relative scientific and independent ecological organism.

For the interior and exterior decoration, bamboo, a traditional architectural material, was used to for the interior ceiling and part of the wall decoration, so as to be more in touch with Chinese culture. (Figure 2-3)

2.2.3 小气候、风水布局与生态整合

风水学是一门反映人与自然生态关系的实用艺术。风水学强调建筑空间的方位布局和环境模型。在当前的生态住宅规划理念中，运用风水理念对小气候进行合理的调节，可以更好地实现宜居的居住环境。

北方印宅贯穿了风水学中的小气候调节概念，使之成为一个独立的自然循环系统。在设计空间流线时，设计了气流的循环路线，利用自然通风和上升气流，结合建筑体块组合的优势，形成理想和谐的生态循环小气候。它提高了生活的舒适度，同时更加自然和绿色。（图2-4）

夏季：采用遮阳措施，避免过多的太阳辐射直接进入建筑中庭，同时利用烟囱效应引导热压通风，风从中庭底部进入，从中庭顶部排出。

冬季：白天充分利用温室效应，夜间采用遮阳措施增加热阻，防止热量散失。

过渡季节：当室内温度过低时，充分利用中庭的烟囱效应带动建筑内各个房间的自然通风，将室内房间和中庭收集的空气带走。（图2-5）

在总体布局上，合理协调和规划建筑主体、半室外空间和室内空间，形成良好的自然通风小气候调节系统和生态有机体。

图2-4 北方印宅夏季内循环图示
Figure 2-4 The Wind Circulation in Summer of Yin House Design Project

2.2.3 Microclimate, Feng Shui Layout and Ecological Integration

Feng Shui is an applied art that illustrates the relationship between human and natural ecology. In Feng Shui, the azimuth layout and environmental model are emphasized. In the current ecological housing planning concept, the rational adjustment of the micro-climate with the concept of Feng Shui can well achieve a livable living environment.

The Yin house runs through the concept of Feng Shui microclimate regulation, making it an independent natural climate circulation system. While designing the spatial streamline, the design stipulates the circulation route of the airflow, utilizing the natural laws of natural ventilation and updraft, and the superiority of the building body block combination forms an ideal and harmonious ecological cycle microclimate. It improves the comfort of living while being more natural and green. (Figure 2-4)

Summer: Use shading measures to avoid excessive solar radiation directly into the building's atrium, while using the chimney effect to guide hot-pressure ventilation, the wind enters from the bottom of the atrium and exits to the top.

Winter: Make full use of the greenhouse effect during the day and use sunshade measures to increase thermal resistance at night to prevent heat loss.

Transition season: When the indoor temperature is too low, make full use of the chimney effect of the atrium to drive the natural ventilation of each room in the building, which is to take away the air collected in the indoor room and the atrium. (Figure 2-5)

In the overall layout, the building main body, semi-outdoor space and indoor space will be rationally coordinated and planned to form a good natural ventilation microclimate regulation system and ecological organism.

图2-5　北方印宅过渡季节内循环图示
Figure 2-5 The Wind Circulation in Transition Seasons of Yin House Design Project

2.2.4 传统形态与平面布局

中国传统建筑中，合院是一种特殊的民居形态，设计师们也尝试从合院的布局出发，力图对传统建筑的形态优势加以创新，用现代手法将起居模式融入到新建筑形体中，以创造符合新中国的居住模式与体验方式。建筑整体平面呈围合式天井院的模式，继承中国传统住宅"四水归堂"的空间特点。（图2-6）

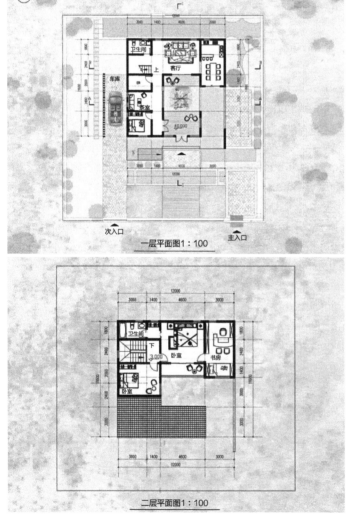

图2-6 北方印宅平面图图示

2.2.4 Traditional Form and Layout

In traditional Chinese architecture, the courtyard is of an unique residential form. The designers also tried to innovate the shape advantages of traditional architecture layout of the courtyard, and integrate it into the new architectural form with modern techniques and living patterns to create a match. New China's living mode and experience. The overall plane of the building is a closed courtyard courtyard model, inheriting the spatial characteristics of the traditional Chinese residence "Sishui Guitang". (Figure 2-6)

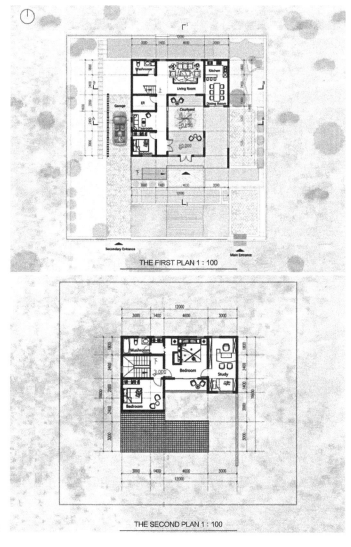

Figure 2-6 Yin House Floor Plan

2.2.5　建筑功能与生态节点

建筑北侧二层，南侧一层，形成一个整体的单坡式屋顶，最大化地提供了太阳能板所需的面积；建筑功能形体为简洁的"L"形，最大化地降低了建筑的形体系数，降低了热能消耗。

在建筑技术上，该项目将模块化建筑、一体化设计、工厂制造、现场施工、验收一体化等技术环节融合在一起。同时，单元体之间可以形成多个空间，合理利用这些空间，可以节约建筑用地。新型建筑材料将使房屋建造和使用更加节能、环保；稻草墙体从制作到应用都更加绿色环保；而太阳能光伏发电装置的设置也降低了能源的消耗，保证室内空气舒适。智能家居是该建筑的另一个主要特点。与普通住宅相比，"北方印宅"不仅具有传统的住宅功能，还具有网络通信、信息家电、设备自动化等功能。水冷空调相比传统空调，制造的空气更加天然、温和。空调系统提供了更舒适的室内生活环境，为居民提供了全方位的信息交互功能，甚至节省了各种能源成本。在墙面处理上，"北方印宅"采用立体绿色外墙面，植物种植在种植槽或容器内，具有吸收粉尘、降低噪声、阻隔有害气体、隔热、节能、挡雨等功能。（图2–7、图2–8）

合院是一个特有的民居形态，我们也尝试从合院的布局出发，力图将传统建筑的形态优势加以创新。

建筑北侧二层南侧一层，并形成一个整体的单坡式屋顶，最大化地提供了太阳能板所需的面积。

建筑功能形体为一简洁的"L"形，最大化降低了建筑的形体系数，进而降低热消耗。

中庭设置自然景观，并将东侧形体打开，使外部景观与中庭景观互相呼应。

置入车库，并在居住者经常活动或休憩的位置分别设置了遮阳设备，以达到遮阳、防辐射的目的。

图2–7　北方印宅建筑功能形体变化图示

2.2.5 Building Function and Ecological Node

The south side of the building is of 1 floor and the north side is of 2 floors, which formed a single sloping roof to maximize the area required for the solar panels. The architectural function is a simple "L"shape that minimizes the shape of the building, reducing heat consumption.

In terms of building technology, modular building, integrated design, factory manufacturing, site construction and acceptance are integrated, and many spaces can be formed between the unit bodies. The rational use of these spaces will save land. New building materials will make the house more energy efficient and environmentally friendly. The straw wall is more green and energy efficient, and the solar photovoltaic setting saves energy and ensures indoor air comfort. Intelligent home is another major feature of the building. Compared with ordinary homes, "Yin House" not only has traditional residential functions, but also has network communication, information appliances, equipment automation, etc. Water-air conditioning is purely natural and friendly. The conditioning system provides a more comfortable indoor living environment, providing residents with a full range of information interaction functions and even saving various energy costs. In the treatment of the wall, the "Yin House" adoped a three-dimensional green wall which is planted in a planting trough or container to absorb dust, reduce noise and harmful gases, while provide heat insulation, save energy and retain rain at the same time. (Figure 2-7, Figure 2-8)

| The courtyard is of unique residential form, and we tried to innovate the advan-tages of the traditional architecture layout. | On the south side to the north side of the building, a single slope roof is formed to maximize the area required for solar panels. | The functional form of the building is a simple "L" type, which maximizes the shape coefficient of the building and re-duces the thermal consumption. | The atrium sets up the natural land-scape and opens the eastern body, echoing the exterior landscape with the atrium. | The shading device is installed in the garage and in the place where the regular activities or rest happen, so as to achieve the purpose of shading & radiation protection. |

Figure 2-7 Functional Forms Diagram of Yin House

花园透视
GARDEN PERSPECTIVE

图2-8　北方印宅建筑外墙效果图
Figure 2-8　Perspective of Exterior Wall of the Yin House

院内透视
COURTYARD PERSPECTIVE

北方印宅设计项目主要使用了变形和累加两种组合方式，是一个将生态环保、地方理念与模块化建筑本身相结合的典型案例。通过对中国传统文化和中国传统民居建筑特色的提炼，在保证生态、节能的前提下保留了中国传统文化元素，并将中国风水文化元素与生态民居的应用相结合，使模块化建筑能够更好地适应农村环境改造。北方印宅设计案例使模块化建筑在建筑领域中的普遍采用成为可能，此建筑建造形式生态、环保，可以在全国各地快速地大规模建设，也为中国乡村改造提供了一种新的设计思路。

2.3 本章小结

模块化是一种设计思维和设计手段。模块的概念不仅存在于建筑设计、计算机、工业技术等领域，也存在于人文、社科、心理等领域。在不同领域，模块都具有通用的属性——灵活性、互换性。

本章节整理并总结了部分在建筑空间领域与模块化相关的理论研究，并选取了一些空间领域的典型案例，用图解的方式诠释了空间模块的特征与模块之间的组织方式。分解、插接、移动、变形、累加、替换、拼合是模块之间的几种组织形式。通过总结这些模块之间的组织形式，可以更好地用模块化建筑设计法进行空间设计。

The Yin House design project is a typical example of combining ecological and local ideas with modular architecture. Through the extraction of Chinese traditional Feng Shui culture and Chinese traditional residential buildings, the traditional Chinese cultural elements are retained on the premise of ensuring ecology and energy saving, and the combination of Chinese Feng Shui cultural elements and the application of ecological residential buildings, so that modular buildings can better adapt to the rural reconstruction environment. Yin House design case makes modular architecture possible for both ecological design and local and rapid large-scale construction. It provides a new design idea for China's rural reconstruction.

2.3 Conclusion

Modularization is a kind of design thinking and means. The concept of modules not only exists in architectural design, computer, industry, technology and other fields, but also in humanities, social sciences, psychology and other fields. Modules have common properties in different domains: they are flexible and can be replaced.

This chapter sorted out and summarized relevant theoretical research on modularization in the field of architectural space, selected some typical cases in the field of space, summarized and illustrated the characteristics of space modules and the way of organization between modules. Separation, connection, movement, transformation, superposition, replacement, and splicing are several organizational forms between modules. By summarizing the organization form between these modules, we can better use modular architectural design method for space design.

3 模块化建筑设计与文化传承

模块化建筑自身具有的灵活性、独立性特点。这些特点为文化继承提供了条件。模块化建筑的独立性让每个单元模块都可以承载独自的文化元素，而灵活性又能保证整体建筑在设计师的统一设计下能够保持文化的一致性。为了让模块化建筑设计法能够更有效地被使用，作者对影响本土文化元素占比的因素及模块的文化承载方式进行了进一步的研究与总结，以期为不同的模块设计环境提供适宜的解决方案。

3.1 影响本土文化元素在模块化建筑中占比的因素

本土文化与当代建筑设计的融合是设计本土化的必经之路。研究影响本土文化元素在设计中的占比因素，能够让模块化建筑体根据项目实际情况，更好地承载本土文化，发挥模块化建筑设计方法的优势。建筑设计本土化不能一味地追求本土文化的融合而迷失设计本身。本土文化元素在当代建筑设计中占比过高，会导致设计作品走入复古或仿制的误区之中，同时，本土文化元素使用过多也会妨碍现代技术的运用，限制设计师设计思维的发挥。而文化元素占比过低，又体现不出建筑设计的本土特色，难以发挥建筑本身的文化承载特性，阻碍设计本土化。本书把影响本土文化元素在当代建筑设计中占比的因素归纳为两点——主观因素与客观因素。

3 Modular Architecture Design and Cultural Inheritance

Modular building itself has the characteristics of flexibility and independence. These character provide conditions for cultural inheritance. The independence of the modular building allows each module unit to carry its own cultural elements, while the flexibility ensures the overall cultural consistency of the building under the unified design of the designer. In order to make the modular architectural design method more effective, the author further studies and summarizes the factors that affect the proportion of local cultural elements in the modular design and the cultural carrying mode of modules, in order to provide appropriate solutions for different module design environments.

3.1 Factors Influencing the Proportion of Local Cultural Elements in Modular Architecture

The integration of local culture and contemporary architectural design is the only way to design localization. The research of factors affecting the proportion of local cultural elements in the design can make the modular building become the carrier of local culture better and make full use of the advantages of modular architectural design method according to actual situation of the project. The localization of architectural design cannot blindly pursue the integration of local culture and lose the design. The excessively high proportion of local cultural elements in contemporary architectural design will lead to the mistake of retro or imitation design works. At the same time, the excessive use of local cultural elements will also hinder the application of modern technology and limit the extension of the design thinking. However, if the proportion of cultural elements is too low, the local characteristics of architectural design cannot be reflected, and it is difficult to give play to the cultural carrying characteristics of architecture itself, which hinders the localization of the design. This book summarizes the factors which influence the proportion of local cultural elements in contemporary architectural design into two aspects: subjective factors and objective factors.

3.1.1 主观因素

建筑设计师在设计过程中既要考虑建筑的未来受众群体又要考虑设计方案委托方的设计诉求。通常委托方的个人主观因素会决定方案的最终结果，而不是建筑设计师。因此，委托方的决策对文化元素的占比起到了关键作用，比如说，个人住宅、商业空间设计等。建筑受众的喜好也是影响建筑设计的主要主观因素之一。每个个体对于本土文化的接受程度不同，为了了解普通大众对我国本土文化的认知和本土文化对生活的影响，本书通过调查问卷的形式收集信息，以此为一项依据来研究中国居民对中国本土文化的认可程度。

问卷调查结果统计：中国居民对中国传统文化接受度 表3-1

	小于或等于30岁	大于30岁	总计
接受	77.6%	83.9%	81.3%
不接受	22.4%	16.1%	18.7%
受访者总数	170人	230人	400人

从表3-1中可以看出，81.3%的受访者（N=400）对在当代建筑设计中加入中国文化元素持认可态度，尽管接受问卷调查的人群可能对中国本土文化的了解并不深入。年龄偏大的受访者对本土文化的接受比例要高于年轻人。受访者给出的接受理由如下：

（1）中国本土文化是中国人几千年来的智慧结晶，有十分重要的历史价值。

（2）中国人对自己的文化一定要有认同感。

（3）现代化的摩登大楼比比皆是，审美疲劳。有中国文化特征的建筑会更有趣。

3.1.1 Subjective Factors

In the design process, architectural designers should consider both the future audience of the building and the design appeal of the client. Sometimes, the client's personal subjective factors will determine the final outcome of the project, not the architect. Therefore, the decision of the client play a key role in the proportion of cultural elements, such as the design of personal homes, commercial spaces, etc. On the other hand, the likes and dislikes of architectural audience are also one of the main subjective factors affecting architectural design. Each individual has different degrees of acceptance for local culture. In orde to understand the public's cognition of China's local culture and the impact of local dialect on life, this book collects information in the form of questionnaire to discover the degree of acception of Chinese local culture by Chinese residents.

Results Statistics of Questionnaire: The acceptability for Traditional Chinese Culture

Table 3-1

	Below or Equal to 30 Years Old	Above 30 Years Old	In Total
Accept	77.6%	83.9%	81.3%
Don't Accept	22.4%	16.1%	18.7%
Total Number of Respondents	170	230	400

It can be seen from Table 3-1 that 81.3% of the respondents (N=400) approve of adding Chinese culture elements into contemporary architectural design. Although the respondents may not have a deep understanding of Chinese local culture. Older respondents accept the local culture more than younger ones. Respondents gave the following reasons for acceptance:

(1) China's native culture is the essence of the wisdom of the Chinese people for thousands of years and has its own value.

(2) Chinese people must have a sense of identity with their important historical value.

(3) Modern design buildings are everywhere and of huge number. It makes aesthetic fatigue to people. Buildings with Chinese cultural features will be more interesting.

通过以上理由可以看出，中国本土文化对中国人的影响是根深蒂固的。因此，中国本土文化也会影响中国人对建筑设计的偏好。尤其是近二十年，在中国城市化的快速发展中，现代化建筑林立，但其往往缺少了中国本土文化元素。为此，笔者也通过问卷的形式调查了一部分普通大众对不同建筑风格的喜爱程度（表3–2）。

问卷调查结果统计：喜爱的建筑设计风格　　表3–2

	小于或等于30岁	大于30岁	总计
传统中式风格	18%	46%	34%
欧式风格	10%	8%	9%
现代中式风格	43%	42%	42.5%
现代风格	16%	4%	9%
其他	13%	0%	5.5%
受访者总数	170人	230人	400人

从表3–2中可以看出，大多数被调查者选择了中国元素与现代相结合的风格，而中国传统风格的认可程度也很高。这一结果表明，中国本土文化元素的使用在当代中国建筑设计中存在需求，并具有广泛的认可度。该问卷调查的结果体现了中国居民对本土文化的主观倾向，证明了将本土文化元素融合到当代建筑设计中的做法具有可行性和发展潜力。同时，中国居民作为中国模块化建筑设计的主要受众群体，其对文化继承的喜好与接受程度不同，也会影响本土文化元素在模块化建筑设计中的占比，体现了主观因素对建筑设计文化继承的影响。

It can be seen from the above reasons that the influence of Chinese native culture on Chinese people is profound. Therefore, local Chinese culture will also influence Chinese people's preference for architectural design. Especially in the past 20 years, with the rapid development of China's urbanization, there are many modern buildings, but most of them lack Chinese local cultural elements. Therefore, the author also investigated the love of some ordinary people for different architecture styles in the form questionnaire (Table 3-2).

Results Statistics of Questionnaire: Favorite Styles of Architecture Design

Table 3-2

	Below or Equal to 30 Years Old	Above 30 Years Old	In Total
Chinese Traditional Style	18%	46%	34%
European Style	10%	8%	9%
Combination of Chinese and Modern Style	43%	42%	42.5%
Modern Style	16%	4%	9%
Others	13%	0%	5.5%
Total Number of Respondents	170	230	400

From Table 3-2, most of the respondents choose a style that combines Chinese elements with modern style. And the recognition of traditional Chinese style is also very high. This result indicates that the use of local Chinese cultural elements in contemporary Chinese architectural design is in demand and widely acceptable. The results of the questionnaire reflect the subjective tendency of Chinese residents towards local culture, which proves the feasibility and development potential of integrating local cultural elements into contemporary architectural design. At the same time, Chinese residents, as the main audience of modular architectural design in China, have different preferences and acceptance of cultural inheritance, which will also affect the proportion of local cultural elements in modular architectural design, reflecting the influence of subjective factors on cultural inheritance of architectural design.

3.1.2 客观因素

除了主观因素外，建筑自身属性所带来的客观因素也对本土化元素在建筑设计中的占比起到了决定性作用。这里提及的客观因素，主要包括：

（1）建筑的使用目的。建筑的使用目的可以分为商业、公益、个人居住等。

（2）建筑的存在时间。建筑存在时间越长，其承载的本土文化元素就越多，本土文化元素的占比也会越高。新建建筑作为文化元素载体的职能相对较弱，强行提高本土文化元素反而适得其反。

（3）建筑模块的生产难度与效率。由于模块化建筑设计以工业化生产为主要环节，因此建筑模块与本土文化元素结合的生产难度与产出效率，成为制约模块化建筑设计的客观因素之一。

（4）建筑的建造成本。模块化建筑的优势之一在于较低的时间与材料成本，也是影响本土化元素在模块化建筑设计中占比的客观因素之一。文化元素的融合会增加一定的模块材料成本和建造时间成本。平衡建造成本与文化继承之间的关系，也是模块化建筑设计需要考虑的客观因素。

基于以上论述，作者选择了建筑设计和本土化文化元素相融合的相关设计项目。这些建筑设计项目的使用目的和客户要求各不相同，因而最终呈现的文化元素占比也有很大差别。下面将分别对北方印宅项目、朝阳街博物馆设计项目、空间改造书店项目、烟台日报社项目，进行详细介绍和横向比较，以此来论证主观因素和客观因素对文化元素占比的影响。

3.1.2 Objective Factors

In addition to the subjective factors, the objective factors brought by the building's own attributes also play a decisive role in the proportion of local elements in architectural design. The objective factors mainly include:

(1) Purpose of the building. The purpose of the building can be divided into commercial purpose, public purpose, personal residence and so on.

(2) The existing time of the building. The longer the building exists, the more local cultural elements it carries, and the higher the proportion of local cultural elements will be. The function of newly built buildings as carriers of cultural elements is relatively weak, forcing to improve the local cultural elements is at cross purpose.

(3) Production difficulty and efficiency of building modules. As modular architectural design takes industrial production as the main link, the production difficulty and output efficiency of combining architectural modules with local cultural elements have become one of the objective factors restricting modular architectural design.

(4) Construction cost of the building. One of the advantages of modular architecture is the lower cost of time and materials, which is also one of the main objective factors affecting the proportion of local elements in modular architectural design. The integration of cultural elements will increase the cost of module materials and construction time. Balancing the relationship between construction cost and cultural inheritance is also an objective factor to be considered in modular architecture design.

Based on the above discussion, the author chosed related design projects integrating architectural design and localized cultural elements. The purpose of use and client requirements of these architectural design projects are different, so the proportion of cultural elements presented is different. The following will be a detailed introduction and comparison of the project of Yin House, the design project of Chaoyang Street Museum, the project of spatial transformation bookstore and the project of Yantai Daily Office. Through the horizontal comparison of different design works, the influence of subjective factors and objective factors on the proportion of cultural elements will be demonstrated briefly.

3.1.3 设计作品分析

（1）北方印宅设计项目（2018年）

■ 设计背景：

北方印宅设计项目来自2018中国国际太阳能十项全能竞赛（简称SDC2018）。由中国国家能源局和美国能源部联合主办，邀请了来自全球10个国家和地区41所高校的22支赛队参与，于2018年在山东省德州市举办。

大赛要求每支赛队以永久性使用为目标，建造一栋建成面积为120～200平方米的单层或双层太阳能住宅。房屋必须配备齐全的日常家用电器及生活设施，并新增电动汽车及充电设备。该项目以为中国乡村改造提供能够广泛应用的经济适用型住房为主要研究目的，进行建筑设计。

■ 设计理念：

在这个项目中，设计师决心追求一种本地化、传统与现代技术相结合的模式，探索并创造一个新的"生态住宅区"。中国历史悠久，深厚的中华民族传统建筑文化和具有本土特色的生活智慧与现代生态哲学相结合，设计师试图以此作为初步尝试。

印宅项目借鉴了四合院的中国传统建筑布局形式。设计师从院落布局入手，试图创新传统建筑的优势，创造一种具有现代技术和生活方式的新建筑风格。屋顶整体最大化了太阳能电池板的面积；紧凑的"L"形平面减少了建筑系统的热量消耗。在风水理论中，强调方位布局和环境模型，并对建筑部分、半室外空间和室外环境进行了合理的总体规划。根据风水布局的主体，建筑主体部分围绕中庭进行穿插，建筑西侧附加车库，室外环境通过建筑体南侧与东侧的半开敞墙体与建筑中庭进行呼应。这四个部分相互穿插，形成了一个很好的生态有机体。（图3-1）

3.1.3 Design Artwork Analysis

(1) Yi House Design Project (2018)

■ Background:

The Yin House design project comes from the 2018 China International Solar Decathlon (SDC2018). Co-sponsored by the National Energy Administration of China and the U.S. Department of Energy, 22 teams from 41 universities, from 10 countries and regions were invited to participate in the event, which was held in Dezhou City, Shandong Province in 2018.

The competition required each team to build a single-storey or double-storey solar-powered house with a completed area of 120-200 square meters, with the goal of permanent use. Houses must be equipped with complete daily household appliances and living facilities, and add electric cars and charging equipment. The main purpose of the research is to provide affordable housing that can be widely used in rural reconstruction in China.

■ Design Concept:

In this program, the designers are determined to pursue a kind combination of localization, traditional and modern technology with a model to explore a new "ecological residential area". China has a long history, there are profound chinese traditional architectural culture of the Chinese nation for chinese local characteristics of living wisdom, and can complement each other with the modern ecological philosophy of the gene, the designers try to make a preliminary attempt in this program.

The Yin House project borrows from the traditional Chinese layout of a courtyard house. Starting from the layout of the courtyard, the designer tries to innovate the advantages of traditional architecture and create a new architectural style with modern technology and lifestyle. The roof maximizes the total area of solar panels; The compact "L" plan reduces the heat consumption of the building system. In Feng Shui theory, the orientation layout and environmental model are emphasized, and reasonable overall planning is made for the building part, semi-outdoor space and outdoor environment. According to the main body of Feng Shui layout, the main part of the building is interspersed around the atrium, with additional garage on the west side of the building. The outdoor environment is echoed by the atrium through the semi-open wall on the south and east sides of the building. These four parts interweave with each other to form a good ecological organism. (Figure 3-1)

北方印宅建筑设计项目是以SDC竞赛要求为标准进行设计的，但是该项目具有建筑受益人群作为参照目标。从客观因素上讲，竞赛对建筑的节能、环保、易搭建等客观属性有明确要求，这也是未来该项目运用到实际乡村改造时必须要具备的客观因素，所以建筑的客观因素对本设计项目具有影响。从主观因素上讲，该设计项目是以乡村改造为前提条件，目标受众已经明确，受众人群的个人主观因素，也是设计师需要考虑的主要因素之一。基于本项目同时具有公益性和商业性的应用属性，主观因素与客观因素对本项目的影响同样重要。其所使用的本土文化元素相较前两个项目比较适中。

图3-1 印宅整体效果图
Figure 3-1 Aerial View rendering of the Yin House

The Yin House architectural design project is designed according to the requirements of SDC competition, but the project has the architectural beneficiaries as the reference target. From the aspect of objective factors, the competition has requirements for the objective properties such as energy saving, environmental friendly and easy construction, which are also the objective conditions that must be met when the project is applied to the actual rural reconstruction project in the future. Therefore, the objective factors have an impact on this design project. From the aspect of subjective factors, the design project is based on rural reconstruction background, and the design target audience has been defined. The individual subjective factors of the target audience are one of the main factors that designers need to consider. As the project has both public welfare and commercial application attributes, subjective and objective factors are equally important. The local cultural elements used in this project are relatively moderate.

（2）朝阳街博物馆设计项目（2017年）

■ 设计背景：

朝阳街位于烟台山前，始建1872年（清同治十一年）。南起北马路，北到海岸街，全长400米。因其为南北走向，又在烟台山之阳，故名朝阳大街，1912年改称朝阳街。1923年该街铺成柏油路面，成为烟台第一条柏油路。烟台政府要求重新设计一个博物馆，将其作为区域的旅游中心，并向当地居民和游客介绍朝阳街的历史。朝阳街博物馆项目主要为免费公益性质。（图3-2）

朝阳街博物馆设计项目依托于政府规划设计项目。烟台市政府将朝阳街的联排建筑进行统一设计整合，因此博物馆的建筑内部空间具有大小不一、形式多样的特点。由于该项目的建筑体构造复杂且给予的建造周期较短，设计师最终采用了模块化建筑设计法来解决不同形态的室内空间，同时达到缩短建造周期的目的。

图3-2 朝阳街现存德式建筑实景照片
Figure 3-2 Photos of the Existing German-style Buildings on Chaoyang Street

(2) Chaoyang Street Museum Design Project (2017)

■ Background:

The Chaoyang Street is located in front of Yantai mountain, which was built in 1872 (Qing Tongzhi eleventh years), with the length of 400 meters. Because of its north–south direction, it was named Chaoyang Street in 1912. In 1923 the street paved asphalt pavement, this is the first asphalt road in Yantai. The local government required to redesign a museum to be a tour center and also to introduce the local history to local residents and tourists. The main purpose of the Chaoyang Street Museum project is for free public benefit. (Figure 3–2)

Chaoyang Street Museum design project relies on the government planning and design project. Yantai municipal government integrated the row buildings of Chaoyang Street in a unified design, so the interior space of the museum has the characteristics of different sizes and forms. Due to the complex structure of the building and the short construction period, modular architecture was adopted to solve the different forms of interior space, while achieving the goal of shortening the construction period.

■ 设计理念:

朝阳街博物馆主要以海、船、浪等海洋元素进行设计,因为开埠文化之路最开始就是外国人通过船舶穿过海洋,来到中国,之后开埠的发展也是通过码头运输等渔业而得以迅速发展。因此,朝阳博物馆的空间很多区域采用折线造型,如通过起伏的木线条造型墙、起伏的吊顶、折线形的展柜等,来表达海浪的形态,灯光大部分采用的也是蓝色灯光。同时,建筑空间也结合一些朝阳街欧式建筑的造型元素来进行设计。建筑材料多为原始材料,如朝阳街的红砖、青砖、蘑菇石、青石板、混凝土、生锈铁板、原木等。原始材料可以更直观地给人一种沧桑感和历史感,可以更好地将游客带入到环境中。建筑造型以现代风格为主,如折线形的顶棚和展柜等,均将原始材料依附在现代造型上,营造出一种历史感的同时,也不乏展现现代气息。(图3-3)

朝阳街博物馆设计项目受烟台市政府部门的直接领导,所以主管部门决策者的个人主观因素对设计的最终呈现起到了一定影响。但朝阳街博物馆设计项目是以朝阳街的旧建筑体为依托,建筑本身就具有深厚的文化底蕴,并且博物馆的未来使用属性是公益性质,建设目的是为了宣传当地开埠文化及历史,所以在这个项目中,主观因素也要服从于客观因素。因此,当地文化元素在项目中所占比例极高,所有的空间结构、设计造型、材料选用都是为了诠释当地文化而存在。

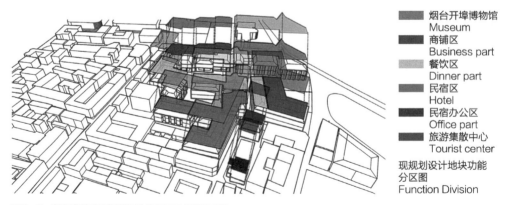

烟台开埠博物馆
Museum
商铺区
Business part
餐饮区
Dinner part
民宿区
Hotel
民宿办公区
Office part
旅游集散中心
Tourist center

现规划设计地块功能分区图
Function Division

图3-3 朝阳街翻新建筑地理位置及功能分区图
Figure 3-3 Function Division of Chaoyang Street

■ Design Concept:

The design concept of this museum mainly comes from the sea, ships and waves and other marine elements. Because the beginning of the port culture is that the foreigners sailed through the ocean by ships and came to the mainland, and then the development of the port makes Yantai city developed more rapidly. Many areas of the space using the polyline shape, such as the wooden wall modeling, undulating ceiling, discounted showcase, etc. to express the meaning of the waves, most of the lights are blue lights. At the same time the design also have some Chaoyang Street European - style architectural elements. Most of the material is the original materials from Chaoyang Street: red bricks, green bricks, mushroom stone, concrete, rusty iron, logs and so on. The use of raw materials, can give a more intuitive sense of the vicissitudes of life; a sense of history, can better bring tourists into the environment. Most of the modeling uses modern styling, the original material attached to the modern style, giving people a sense of history and at the same time, do not lack of modern flavor. (Figure 3-3)

The design project of Chaoyang Street Museum was directly led by Yantai Municipal Government, so the personal subjective factors of decision-makers in charge of the department had a certain influence on the final presentation of the design. However, the design project of Chaoyang Street Museum is based on the old building style of the Chaoyang Street, the building itself is a profound cultural heritage, and the future use of the museum is for public welfare, the purpose of the construction is to publicize the local culture and history of opening port. So in this project, subjective factors should also be subordinate to objective factors. As a result, local cultural elements account for a high proportion of the project, and all the spatial structures, design forms and materials are chosen to interpret the local culture.

（3）空间改造书店设计项目（2019年）

■ 设计背景：

该设计项目位于烟台市中心的万达商业街。客户要求设计师团队将餐厅重新设计成一个商业书店，在室内空间为顾客提供良好的阅读体验。客户想把这家书店经营成集休闲娱乐、体验式阅读、餐饮为一体的商业化空间。以阅读舒适度为盈利手段，通过引导消费者消费的方式兼顾书籍售卖。空间中要融入中国传统文化中与书相关的元素。

设计师为了充分利用空间，提高商业书店的盈利效率，进行了空间模块置入。同时，结合模块的书籍类型，将模块内部与对应文化元素进行结合。模块外部与整体空间应用的中国传统文化元素相统一。

■ 设计理念：

这家书店的设计理念主要来源于中国传统的书籍制作材料和读者的阅读习惯。通过在室内空间添加书籍相关元素，使顾客在室内空间产生阅读的欲望和思想的共鸣。空间中有许多由木头和竹子制成的模型，这些模型也增添了阅读空间的自然感受和舒适度。与此同时，根据不同的阅读需求，也有不同风格的空间来丰富体验，如科幻阅读区的宇宙穹顶和繁星（图3-4）。

因为书籍的种类繁多，为了兼顾客户的诉求，用模块化设计方法在空间进行了模块置入，将科幻区单独间隔出来，既保持了整体的传统氛围不被破坏，又能照顾到爱好科幻读物的顾客的需求。

图3-4　通道视觉效果图
Figure 3-4　Passageway View

(3) Space Transformation Bookstore Design Project (2019)

■ Background:

This design project is located at Wanda Business Street in the center of Yantai City.The clients required designers' team to redesign a restaurant to be a book store that offers a good experience of reading in the interior. The client wants to operate this bookstore into a commercial space integrating leisure and entertainment, experiential reading and catering. With reading comfort as a means of profit, by guiding consumers to consumers, book sales sold simul cutaneously at the same time. The space should incorporate elements related to books in traditional Chinese culture.

In order to make full use of the space and improve the profit efficiency of the commercial bookstore, the designer carried out space module placement. At the same time, combining the book types of the module, the module is combined with the corresponding cultural elements. The exterior of the module is unified with the traditional Chinese cultural elements of the overall space application.

■ Design Concept:

The design concept of this book store mainly comes from the Chinese traditional reading materials and habits. With adding these elements in this interior space, this interior space Will let the customer have the desire to read and the resonance of thought. There are many modelings made by wood and bamboo. These models also increase the natural feeling and comfort of reading. Based on different desire of reading requirement, there also have different elements from the space to enrich the experience, such as the universe and the stars(Figure 3–4).

Because there are different kinds of books in order to satisfy the demands of customers, modular design method is used for placing modules in the space to separate a science fiction reading area, which not only keeps the overall traditional atmosphere from being destroyed, but also takes care of the needs of customers who love science fiction books.

如图3-5所示，框选部分为设计师在书店北侧置入的空间模块，将整个空间进行了分隔，独立的模块空间内部应用了书籍主题文化元素进行设计，外立面搭配阶梯书架和原木材质贴片，与整个书店的传统文化主题相呼应。

空间书店设计项目改造是以客户的经营模式和商业需求为主要导向的，书店店主的个人主观因素对设计的最终呈现起决定性作用。书店设计项目的使用属性是商业性质，建设目的是以商业盈利为主要诉求，所以在这个项目中，书店潜在顾客的主观因素也被客户和设计者考虑在内。在这个项目中，传统文化元素的占比是客户根据未来市场主要受众群体的喜好来决定的。因此，建筑的客观因素被弱化，人的主观因素占据了主导作用。与朝阳街博物馆设计项目相比，商业书店设计项目中的本土文化元素占比相对较少。虽然同为模块化设计项目，但是因为两个项目中主观因素和客观因素的主导作用不同，导致空间模块所承载的本土文化元素占比不同。这种差异性体现了主观因素与客观因素对模块化建筑的本土化文化承载应用的影响。

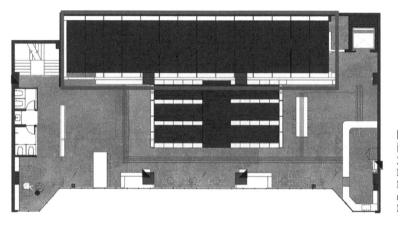

图3-5 书店一层平面：空间模块置入示意图
Figure 3-5 First Floor Plan of Bookstore: Schematic Diagram of Space Module Placement

As shown in Figure 3-5, the selected part is the space module placed by the designer in the north side of the bookstore, which separates the whole space. The independent module space is designed with book theme cultural elements. The exterior facade is matched with stepped bookshelves and log material stickers, which echoes the traditional cultural theme of the whole bookstore.

The space transformation bookstore design project is determined by the customer's business model and business needs, and the personal subjective factors of the bookstore owner play a decisive role in the final presentation of the design. The use attribute of the bookstore design project is commercial and the main construction purpose is to make commercial profit and commercial appeal. So in this project, the subjective factors of the potential customers of the bookstore are also taken into account by the clients and designers. In this project, the proportion of traditional cultural elements is determined by the client based on the preferences of the major audience groups. Therefore, the objective factor of architecture is weakened, and the subjective factor of human occupies became the dominant role. Compared with the Chaoyang Street Museum design project, the local cultural elements in the commercial bookstore design project are relatively less. Although both projects are modular design projects, the dominant role of subjective and objective factors in the two projects is different, resulting in different proportions of local cultural elements carried by the space modules. This difference reflects the influence of subjective and objective factors on the application of localized cultural bearing of modular architecture.

（4）烟台日报社办公空间设计项目
（2019—2020年）

■ 设计背景：

烟台日报社成立于1995年，创立至今已有25年历史。日报社的办公区域坐落于烟台市中心区，临近海港，是一家蓬勃活力的现代化传媒集团。因为老办公区不合理的分区和环境、设施陈旧已经不能适应日报社现在的办公需求，所以根据客户要求，本设计项目主要针对日报社新闻部门办公区域进行重新设计与规划。在满足现代化办公需求的基础上，融合日报社自身的企业文化特色与历史，设计出轻松、舒适、高效、有企业文化内涵的办公空间。

■ 设计理念：

烟台日报社办公空间的设计灵感来自烟台的海洋地貌，因为烟台是著名的海港城市，日报社办公大楼又临近海港，能够远眺大海景色，所以决定在办公空间融入具有地域特色的海洋元素。针对集团对办公空间的灵活性需求，将不同尺度的模块化空间嵌入空间，创造出独立的三个盒子空间——生态盒子、文化盒子、艺术盒子。三个盒子的位置能够使员工从办公区的任意角度看到至少一个盒子创造出的景色或文化，营造轻松、舒适的办公环境。（图3-6）

(4) Yantai Daily Office Design Project (2020)

■ Background:

Founded in 1995, Yantai Daily has a history of 25 years. The office area of Yantai Daily is located in the downtown area of Yantai, near the harbor. It is a modern media group with a relative long history and vigorous vitality. Since the old office area is not properly partitioned and the old office area can no longer meet the current office needs of the daily newspaper, this design project is mainly designed and planned for the office area of the news department of the daily newspaper according to the requirements of the customer. On the basis of satisfying the needs of modern office, integrating the characteristics and history of the daily's own corporate culture, there designed a relaxed, comfortable, efficient office space with corporate culture connotation.

■ Design Concept:

The design of the office space of Yantai Daily is inspired by the marine landform of Yantai. Because Yantai is a famous seaport city, the office building of Yantai Daily is close to the seaport and can overlook the sea scenery from afar, so the Marine elements with regional characteristics are integrated into the office space. According to the company's demand for flexibility office space, modular spaces of different scales are embedded into the space to create three independent box Spaces—ecological box, cultural box and art box. The location of the three boxes enables employees to enjoy the scenery or culture created by at least one box from any angle of the office area, creating a relaxed and comfortable office environment. (Figure 3-6)

前台
Reception

休闲区
Recreational Area

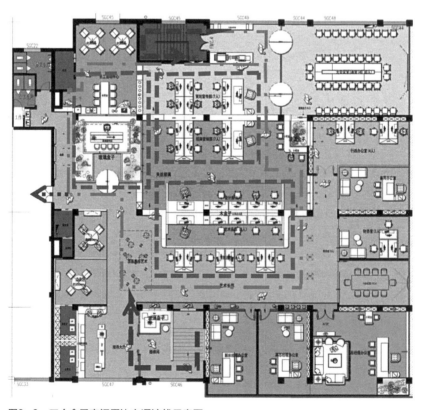

图3-6　三个盒子空间周边交通流线示意图

艺术走廊
Art Passageway

文化盒子办公区
Art Module Office Area

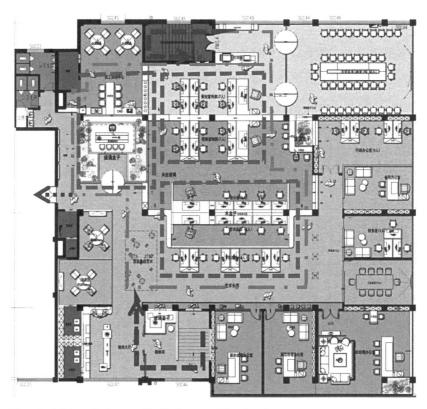

Figure 3-6 Schematic diagram of traffic flow around three box Spaces

　　烟台日报社办公室设计项目属于半封闭式办公环境，以工作交流的使用目的为主，主要受众群体是烟台日报社的工作人员和前来洽谈的外来人士。由于办公空间并不是一个展厅，它的主要功能是在工作中与员工交流，所以企业的相关文化元素与设计项目的融合不会在空间的每一处都有所体现，而是以空间模块为载体集中展现企业文化的方式，将本土文化元素贯穿到整个设计当中。基于烟台日报社办公空间的客观地理位置，滨海环境的客观因素导致该项目中同时提取与融合了海洋元素与企业文化元素，因此客观因素影响了本土文化元素在空间模块中的应用占比。综上所述，在该设计项目中，受主客观因素的影响，文化元素的选择以企业文化为主，以本土文化元素为辅。与其他项目相比，当地文化元素的所占比例适中。

Yantai Daily office design project belongs to a semi-closed office environment, mainly for work and communication, the main audience group is the staff of the Yantai daily and the non-native people who come to negotiate. However, the office space is not an exhibition hall, and its main function is for employees to communicate at work, So the integration of the relevant cultural elements of the enterprise and the design project will not be reflected in every place of the space, but in the way of focusing on the space module to show the corporate culture, the local cultural elements throughout the whole design. Based on the objective geographical location of Yantai Daily office space, the objective factors of coastal environment led to the extraction and integration of marine elements and corporate culture elements in the project, so the objective factors affected the application proportion of local cultural elements in the space module.To sum up, the design of the project was influenced by both subjective and objective factors, and the selection of cultural elements is mainly based on company culture, supplemented by local cultural elements. The proportion of local cultural elements is moderate compared with other projects.

　　图3-7为本土文化元素在四个不同的设计项目中的占比例示意图。元素占比高低的衡量标准以本土文化元素在整个设计表面覆盖面积作为测算依据。齿轮和心形图案分别代表客观因素和主观因素。齿轮和心形图案的数量越多，则代表客观因素或主观因素对该设计项目的影响越大。通过对每个设计项目的主观因素和客观因素的主导作用分析，将项目中本土文化元素的覆盖面积作为基数，分别计算主观因素和客观因素起主导作用的本土化元素覆盖面积，继而得出影响占比数据。

　　通过四个项目的横向比较可以发现，客观因素和主观因素对本土文化元素在模块化建筑设计中的占比高低确实起到了影响。因此，模块化建筑设计的本土化应用需要对项目的主观和客观因素进行调研，以期获得适应性的设计策略。但是需要注意的是，并不是客观因素和主观因素的影响越大，本土文化占比就一定越多。本土文化元素的最终所占比例还会根据实际情况和设计方案上下有所浮动。所以，设计师在使用模块化建筑设计方法进行设计时，要充分考量客观因素和主观因素，将空间模块与本土文化元素结合的应用覆盖面积进行测算，从而设计出既让客户满意，又满足实用性的模块化建筑设计作品。

The Figure 3-7 above diagram shows the proportion of local cultural elements in four different design projects. The measurement standard of element proportion is based on the area covered by local cultural elements on the whole design surface. The gear pattern and the heart pattern respectively represent objective and subjective factors. The greater the number of gears or hearts, the greater the influence of objective or subjective factors on the design project. By analyzing the dominant role of subjective and objective factors in each design project, the coverage area of local cultural elements in the project is taken as the base, and the coverage area of localized elements dominated by subjective and objective factors is calculated respectively, and then the influence proportion data is obtained.

Through the comparison of the four design projects, it can be found that objective factors and subjective factors do have an impact on the proportion of local cultural elements in modular architectural design. Therefore, the localization application of modular architectural design needs to investigate the subjective and objective factors of the project, in order to obtain adaptive design strategies. However, it should be noted that it doesn't mean the greater the influence of objective factors and subjective factors are, the more proportion of local culture is. The final proportion of local cultural elements will fluctuate according to the actual situation and the design scheme. Therefore, when designers use modular architectural design methods to design, they should fully consider objective and subjective factors, and measure the application coverage area of space modules combined with local cultural elements, so as to design modular architectural design works that both satisfy customers and meet practicality.

设计项目分析结论

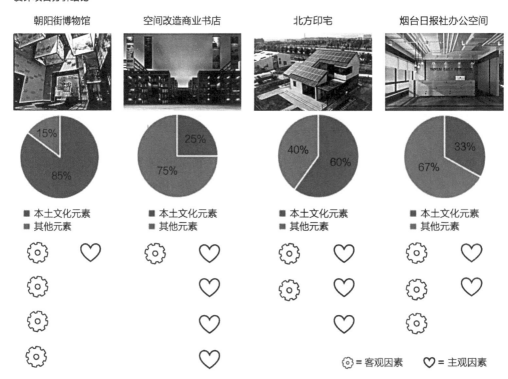

图3-7　项目横向比较分析图

Results of Design Analysis

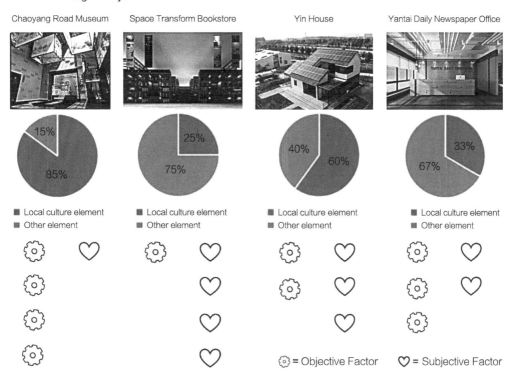

Figure 3 – 7 Horizontal Comparison Diagram of the 4 Design Projects

3.2 模块化建筑设计法的文化承载方式

如今,传统的模块化建筑设计法不足以满足多元化的建筑设计需求。因此,模块化建筑设计法应兼顾建筑体模块构建与室内空间模块构建,其本土化应用策略应将主观因素与客观因素纳入设计过程的考量之中。如此,模块化建筑设计法既继承了传统模块化建筑的灵活性、高效率施工等优点,又具备能够满足设计师设计需求的高配合度。模块化建筑设计法在文化的承载方式上具有多样性,主要分为造型设计、结构布局等客观视觉感知方式进行文化内涵承载,还有空间氛围营造等主观感受方式。

基于模块化建筑的个体独立特性,本土文化元素既可以以单一模块为载体,也可以通过组成完整建筑体成为文化元素载体。模块化建筑的灵活性使其更易于梳理和整合设计项目中的本土文化元素。本土文化元素的占比也处于可控的情况下,可以由模块的数量和表面积计算得出数据,把主观的文化感知转化为客观量化数据,从而使建筑空间和本土文化元素能够更好地融合。这也是模块化建筑设计法在文化传承方面的优势。

3.2 Cultural Carrying Mode of Modular Architectural Design Method

Nowadays, the traditional modular architectural design method is not enough to meet the diversified architectural design needs. Therefore, modular architectural design method should take into account both building module construction and interior space module construction, and its localization application strategy should take subjective and objective factors into consideration in the design process. In this way, modular building design method not only inherits the advantages of flexibility and high efficiency construction of traditional modular buildings, but also has the high degree of cooperation to meet the design needs of designers. The modular building design method proposed in this research inherits the advantages of the traditional modular building, such as flexibility and high construction efficiency. At the same time, this method has high adaptibility to satisfy the design needs. Modular architectural design method has diversity in the way of carrying culture. It not only uses objective visual perception ways such as modeling design and structural layout to carry out cultural connotation, but also uses subjective feeling ways such as space atmosphere creation.

Based on the individual independence of modular architecture, local cultural elements can be either a single module as a carrier or a whole building body as a carrier of cultural elements. The flexibility of the modular building makes it easier to tease and integrate the local cultural elements into the design project. The proportion of local cultural elements is also under control by this design method. The subjective cultural perception can be transformed into objective quantitative data by calculating the number and volume of modules which have cultural elements. So that the architectural space and local cultural elements can be better integrated. This is the advantages of modular architectural design method in terms of cultural inheritance.

3.2.1 北方印宅建筑设计项目

北方印宅建筑设计项目是从建筑空间的大尺度上通过结构布局来承载本土文化元素的典型案例。以中国徽派建筑外观为主要参照、中国传统四合院格局为主要建筑布局形式来设计的印宅，每个单一模块和组成的建筑整体都能发挥其文化承载职能。因为单一模块具有独自承载文化元素的能力，徽派建筑的灰瓦白墙在每一个模块都能独立呈现，不受整体模块数量和摆放位置的影响。因此，模块化建筑设计法在文化承载方面具有灵活性、个体独立性和整体性，这有利于本土文化元素与当代建筑设计相融合的实现。

建筑整体规划为四合院环绕格局，继承了中国传统民居"四水为中心"的空间特征。中庭设置自然景观，与东侧主体开放，外部景观与中庭场景相互呼应。主楼北侧分为两层，南侧一层为入口，北高南低，具有中国传统建筑形态特色（图3-8）。

客厅布置在一层北侧的中间位置，厨房、浴室两侧分别布置于房屋的东北方向和西北方向。房屋的西侧和东侧，布置了卧室和餐厅，无论身处哪个房间，都可以享受庭院空间的环境。在一楼穿插室内外两个艺术展区和一个贴近自然的生活空间。二楼主要设置卧室和书房空间，以满足房间的采光需求。

3.2.1 Masterwork: Yin House Architectural Design Project

Yin House architectural design project is a typical design case which carrying local cultural elements through structural layout on a large scale of architectural space. With the appearance of Chinese Hui—style architecture as the main reference and the pattern of Chinese traditional courtyard as the main architectural layout, each single module and the whole building can play its cultural bearing function. Because a single module has the ability to carry cultural elements independently, the gray tiles and white walls of Hui–style architecture can be presented independently in each module. It is not affected by the number and placement of the whole modules. Therefore, modular architectural design method has flexibility, individual independence and integrity in terms of cultural bearing, which is a favorable condition for the integration of local cultural elements and contemporary architectural design.

The overall plan of the building is surrounded by courtyard pattern, which inherits the spatial characteristics of Chinese traditional residential "four water to the center". Natural landscape was set up in the atrium, and the east side of the main body is open up, the external landscape and atrium scene echoed each other. The north side of the main building is divided into two layers, one layer is formed on the south side which is the entrance, and the north south high low form, is complied to China traditional architectural form(Figure 3–8).

The living room is arranged in the middle position on the north side of the main layer, the kitchen and bathroom is arranged on the northeast and northwest sides of; on the west side and the east side, is the bedroom and dining room where can enjoy the courtyard space; there also interspersed with two indoor and outdoor art exhibition area on the first floor, which is a livable space close to the nature; the bedroom and study space were mainly set up on the second floor, to meet the lighting requirements.

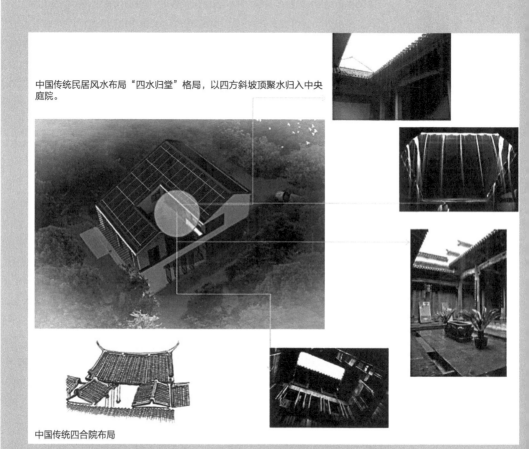

中国传统民居风水布局"四水归堂"格局，以四方斜坡顶聚水归入中央庭院。

中国传统四合院布局

图3-8 北方印宅的中国传统建筑元素提取分析图

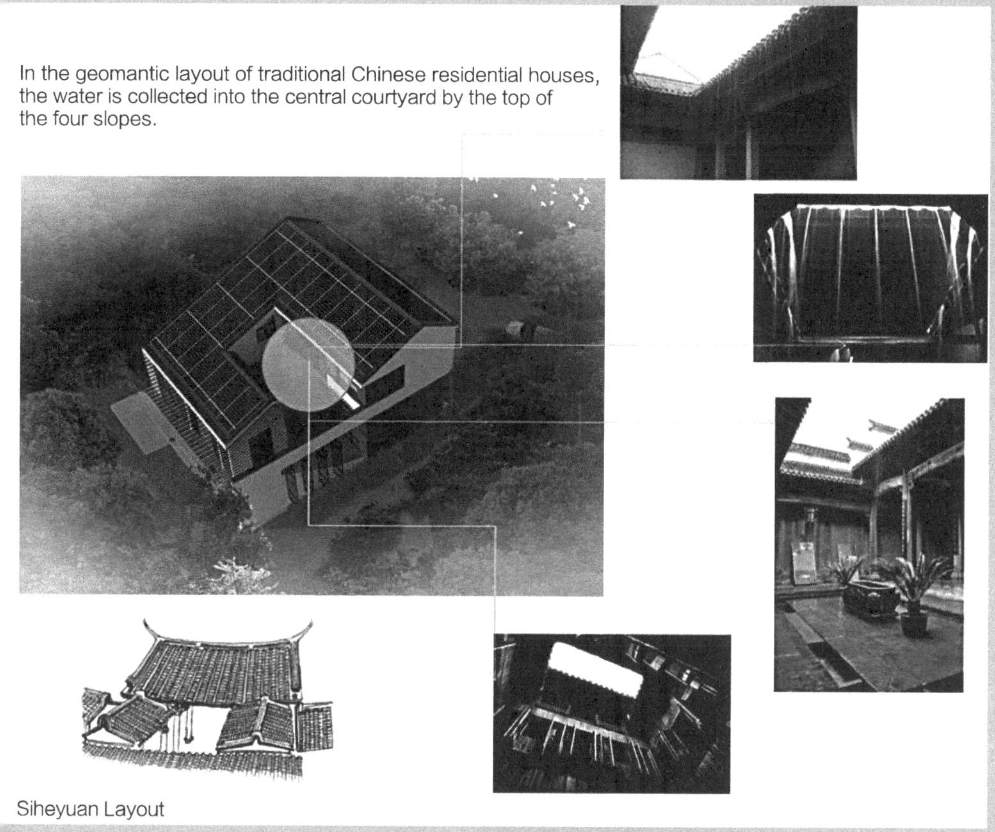

In the geomantic layout of traditional Chinese residential houses, the water is collected into the central courtyard by the top of the four slopes.

Siheyuan Layout

Figure 3–8 Chinese Traditional Architectural Elements Extraction and Analysis Diagram

　　从文化承载的角度来说，印宅项目不仅继承了中国传统徽派建筑的主要特征（白墙灰瓦）和四合院的平面布局，还通过在中庭设置自然景观，东侧形体打开，外部景观与中庭景相互呼应，体现了"天人合一"的中国传统思想。房屋顶部用太阳能板替换掉灰瓦，将节能技术与中国传统文化元素结合在一起。作为本土文化元素的建筑载体，这些特征的呈现效果不会因为模块数量和大小的变化而有很大的出入。相反地，因为提取的本土文化元素，如四合院的围合方式、徽派建筑的建筑外貌等特征，建造工序简易且具有中国传统民居代表性，沿用了传统模块化建筑的工业化产品，没有因为本土文化的融合而改变整体的工业生产流程。因此，在北方印宅项目中，中国传统文化元素与现代技术的融合获得了实际的经济效益。（图3–9）

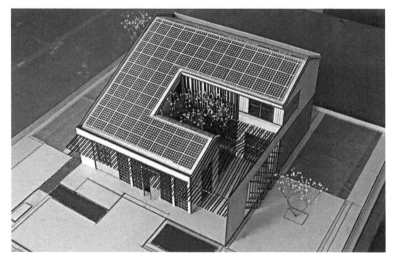

图3–9　北方印宅模型照片
Figure 3–9 Model of Yin House

From the perspective of cultural bearing, the Yin House project not only inherits the main features of Chinese traditional Hui-style architecture (white walls and grey tiles) and the layout of the quadrangle courtyard, but also reflects the traditional Chinese thought of "harmony between nature and man" by setting up natural landscape in the atrium.The roof of the house is replaced gray tiles with solar panels, which combine energy-saving technology with traditional Chinese cultural elements. As the architectural carrier of local cultural elements, the presentation effect of these features will not vary greatly with the change of the number and size of modules. On the contrary, because of the extracted local cultural elements, such as the enclosing way of siheyuan and the architectural appearance of Hui-style buildings, the construction process is simple and representative of traditional Chinese residential houses, and the industrial production products of traditional modular buildings are used, without changing the overall industrial production process due to the integration of local culture. Therefore, in the Yin House project, the integration of traditional Chinese cultural elements and modern technology has achieved practical economic benefits. (Figure 3-9)

3.2.2 朝阳街博物馆设计方案

在朝阳街博物馆案例中，每个展厅被设计成不同的历史文化主题，以历史事件发生进程的时间轴为线索，将各个展厅模块串联在一起，使该博物馆成为朝阳街历史文化的载体，对当地历史文化进行了完整的阐述。

朝阳街博物馆的大厅设计主要以模块墙体的文化造型为文化承载对象。置入模块的三面墙体都使用了木质的海浪起伏造型墙，暗示开埠文化与海洋之间的关联。屋顶悬挂四个中空小模块，展示了朝阳街的老旧照片。通过视觉的直观感受和造型墙的暗示，从入口处就把参观者带入开埠文化的情境之中。（图3–10）

朝阳街博物馆作为朝阳街历史文化的叙述载体，其文化元素的全面展现是设计师首要考虑的问题。于是，设计师为项目中的每一个展厅都赋予了关于朝阳街文化的主题。每一个展厅模块都承载着朝阳街开埠文化的一部分。（图3–11）

序厅的船帆造型和码头模拟都是以文化造型设计为主要文化承载方式。通过这些造型元素来展现开埠是始于外来侵略者抵达烟台港口的文化。序厅通道呈现出空间逐渐打开的造型，从视觉和空间氛围上暗示参观者开埠文化由此开始。

3.2.2 Design Scheme of Chaoyang Street Museum

In the design case of Chaoyang Street Museum, each exhibition hall is designed with different historical and cultural themes. With the timeline of historical events as the clues, each exhibition hall module is connected together, making the museum to be the carrier of the history and culture of Chaoyang Street and carrying out a complete exposition of the local history and culture.

The design of museum reception hall mainly takes the cultural shape of modular wall as the cultural bearing object. Wooden wave-shaped walls are used as wall on three sides of the modules, suggesting the connection between the port culture and the sea. Four small hollow modules hang from the roof, showing old photos of Chaoyang Street. Through the visual intuition and the suggestion of the shaped wall, visitors would be brought into the context of the opening port culture from the entrance. (Figure 3-10)

The Chaoyang Street Museum as a narrative carrier of Chaoyang Street history and culture, the comprehensive display of its cultural elements is the designer's primary consideration. Therefore, every exhibition hall in the project has been given a theme about the culture of Chaoyang Street. Each module of the exhibition hall carries a part of the opening port culture of Chaoyang Street. (Figure 3-11)

The "sail-like" modeling of the introduction exhibition hall and the dock simulation of preface exhibition hall are both based on cultural modeling design as the main cultural bearing mode. Through these modeling elements, it shows that the opening port culture started from the arrival of foreign colonists to Yantai port. The passageway of the preface exhibition presents the shape of space than gradually open. The visual and spatial atmosphere suggests that the opening port culture begins from here.

图3-10 接待大厅

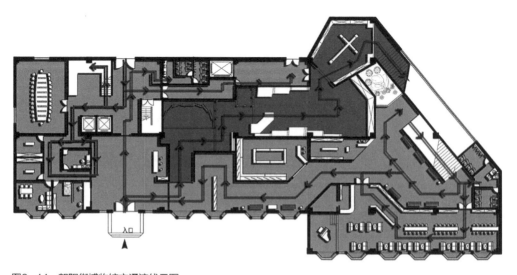

图3-11 朝阳街博物馆交通流线示图

Figure 3 – 10 Reception Hall

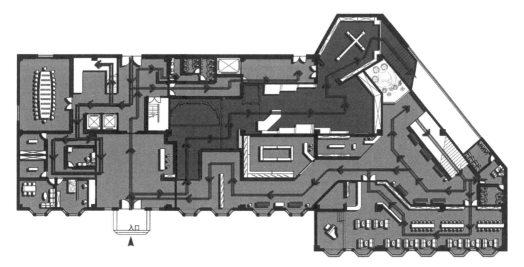

Figure 3 – 11 Traffic flow diagram of Chaoyang Street Museum

基督教文化展厅和历史进程展厅都是通过模块的空间结构布局进行文化传承。基督教文化展厅为了接引自然天光进入空间形成"光"十字架，融合两层楼体，将基督教文化元素传入朝阳街的历史，通过小空间的构造变化展现出来，同时营造了基督教展厅的朝圣氛围。历史进程展厅则通过两侧墙壁形成的喇叭形状，将主要的历史事件展示在两侧墙壁和顶棚上。以广角镜的构造最大限度地扩大参观者的视角，以期在最短时间内将大量历史信息传达给参观者。

张裕葡萄酒公司展厅和纪念品商铺使用了朝阳街现存建筑的红砖和灰砖元素，以期还原当时的街景，让参观者感受朝阳街的历史文化氛围。这也体现了模块化建筑呈现的空间氛围营造方式对文化的承载（图3-12）。

图3-12　纪念品售卖区
Figure 3-12　Souvenir Shop

The Christian introduction exhibition hall and the historical process exhibition hall carry culture through the cultural bearing mode of spatial structure layout. In order to bring natural light into the space and form a cross of light, the Christian Culture Exhibition Hall integrates a two—storey building, and shows the history of Christian cultural elements into Chaoyang Street through the structural changes of the small space, and meanwhile creates the sacred atmosphere of the Christian introduction exhibition hall. The historical process exhibition hall displays the main historical events on the walls and ceiling through the flared shape of the two walls. The structure of wide–angle structure maximizes the viewing angle of visitors, in order to be able to transmit a large amount of historical information to visitors in the shortest time.

The Zhangyu Wine exhibition hall and the souvenir shop use red and gray bricks which come from the existing buildings on Chaoyang Street, in order to restore the street scene at that time and give visitors a sense of the historical and cultural atmosphere of Chaoyang Street. This also reflects the subjective culture carrying mode of the space atmosphere of modular building. (Figure 3–12)

3.2.3　商业书店设计项目

　　商业书店设计项目在客户的要求下，以中国传统文化元素为主要设计元素。基于书籍的种类众多，中国传统文化元素无法满足所有读者的阅读兴趣，所以整个书店室内空间的中国传统文化元素被设计师弱化，尽可能让不同种类的阅读区能彼此协调。不同于朝阳街博物馆文化元素的大面积使用，商业书店项目以点带面，用局部传统文化造型带动整体氛围，发挥模块化建筑的空间氛围营造和造型设计的文化承载方式。而科幻主题阅读区因为独立在整体环境之外，虽然使用的文化承载方式与整体一致，但是因为大面积科幻元素的应用，科幻主题氛围直观且浓烈，与整体的设计表现技巧上有本质区别。由此可见，同样的文化承载方式结合不同的设计表现方法，得到的最终效果也不尽相同。

　　阶梯书架造型使用了棕色原木作为饰面。通过大面积木质材料的使用来确定整个书店空间中央区域的传统文化氛围，以中央区的突出位置和贯通上下两层的台阶让阶梯书架造型成为空间焦点。这样以点带面，既能平衡书架之间的六个不同种类阅读区的设计风格，又能与整体设计风格相呼应，营造中国传统文化氛围。（图3-13）

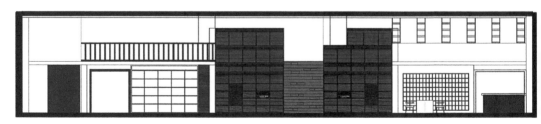

图3-13　阶梯书架立面图示
Figure 3-13 Stair Book Shelves Elevation

3.2.3 Commercial Bookstore Design Project

Under the requirements of customers, commercial bookstore design project uses the Chinese traditional cultural elements as the main design elements. Due to the large variety of books, conventional traditional Chinese cultural elements cannot satisfy the reading interests of all readers. Therefore, the traditional Chinese cultural elements in the interior space of the bookstore are weakened by the designer, so that different types of reading areas can coordinate with each other as much as possible. Different from the large-scale use of cultural elements in Chaoyang Street Museum, the commercial bookstore project replaces the whole surface with small design points, which drives the overall atmosphere with local traditional cultural modeling, and gives full play to the space atmosphere created through modular building design method, and the cultural bearing mode of the modeling design. At the same time, the science-fiction theme reading area is independent from the overall environment. Although the used cultural carrying mode is the same as the overall environment, the science-fiction theme atmosphere is strong due to the application of large area of science-fiction elements, which is essentially different from the overall design performance skills. Thus, the same cultural carrying mode combined with different design expression methods will produce different final effects.

Brown logs were used as the veneer for the stepped bookcase. The traditional cultural atmosphere in the central area of the whole bookstore space is determined, through the use of large area of wood materials. The prominent location of the central area become the focus of the space through the two layers of steps which are the staircase shaped bookshelf. In this way, it can balance the design style of six different types of reading areas between the bookshelves structure, and resonance with the overall design style to create a traditional Chinese cultural atmosphere. (Figure 3-13)

在阶梯书架对面，设计师参照中国古代竹简书的样式设计了两个竹简书造型悬挂在墙壁上。在纸张还没有发明之前，中国一直以竹简作为主要书籍材料。所以，竹简造型与中国传统文化和书籍都有密不可分的关系，是具有辨识度的典型文化载体。把竹简造型放在这个位置，可以增强阶梯书架的中央区影响，用小空间烘托整体的中国传统文化氛围。（图3-14、图3-15）

为了平衡不同顾客的阅读爱好，二楼阅读区设计的比较普通，同样是以点带面的设计手法，设计师在阅读区的墙壁上悬挂了一些带年轮的树墩制作的装饰灯具，为整个阅读区增加原木元素，与空间整体的设计风格相呼应。（图3-16、图3-17）

图3-14　竹简书造型效果图
Figure 3-14 Bamboo Slips Book Modelling Rendering

On the opposite side of the stepped bookshelves, two shapes are designed with reference to the style of bamboo slips in ancient China to be hung on the wall. Before paper was invented, the bamboo slips were used as the main writing carrier in China. Therefore, the shape of bamboo slips is closely related to Chinese traditional culture and books, and it is a typical cultural carrier with recognition. To put the bamboo slip modeling in this position can enhance the influence of the central area with the stepped bookshelf. This is a use of small space as a design point to foil overall Chinese traditional culture atmosphere. (Figure 3 – 14, Figure 3 – 15)

In order to balance different reading hobbies of customers, the design of reading area on the second floor is relatively common. The same design concept is applied, which is to replace the whole surface design with the small design point. The designer hangs some decorative lamps made of tree stumps with tree rings on the wall in the reading area, adding log elements to the whole reading area to echo the overall design style of the space. (Figure 3 – 16, Figure 3 – 17)

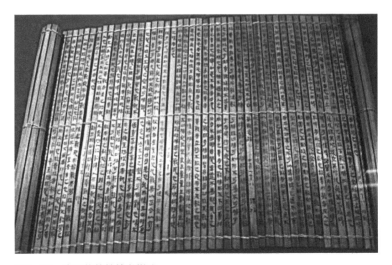

图3–15 中国传统竹简书样式
Figure 3 – 15 Chinese Traditional Bamboo Slips

图3-16　二楼阅读区
Figure 3 – 16 The Second Floor Reading Area Rendering

图3-17　二楼阅读区立面图示
Figure 3-17 The Second Floor Reading Area Elevation

3.2.4　烟台日报社办公空间设计项目

烟台日报社办公空间设计项目以结构布局为主要文化承载方式，同时辅以造型设计。通过三个模块的置入，尤其是文化模块的半开敞造型和对功能区的分隔，使员工的日常使用中潜移默化地将企业文化中的协作精神深入每个员工的心中。本方案借助模块的结构布局，营造出团结协作的企业文化氛围，这也是模块化建筑设计法中主观文化承载方式的实际应用。

由烟台日报社的设计元素分析图可以看出，空间中的主要文化元素都是基于企业文化和临近的海洋地理位置这两点来进行文化元素的提取和转化的。为了更好地体现烟台日报社企业文化中的团结协作精神，设计师将文化模块置入办公空间的中央位置，将其内部设计成协作讨论区。开敞的环境让模块周围的四个办公区域都能够自由使用讨论区，从而增加不同部门之间的协作。这里使用了文化承载方式中的空间氛围营造，利用模块化设计推动人与人之间的交流协作。（图3－18）

3.2.4 Yantai Daily Office Design Project

The Yantai Daily Office Design Project takes the structural layout as the main cultural carrying mode, supplemented by the cultural carrying mode of cultural modeling design. Through the placement of three modules, especially the semi-open shape of the culture module and the separation of functional areas, the spirit of collaboration in the corporate culture can be imperceptibly embedded into the thoughts of every employee in their daily use. With the help of the structural layout of modules, this design project creates a corporate culture atmosphere of solidarity and cooperation, which is also the practical application of the subjective culture bearing mode in modular architectural design method.

It can be seen from the analysis diagram of design elements that the main cultural elements in the space are extracted and transformed based on the corporate culture and the adjacent marine geographical location. In order to better reflect the spirit of unity and cooperation in the corporate culture of Yantai Daily, the designer placed the cultural module in the central position of the office, and designed it as the collaborative discussion area. The open environment allows the four office areas around the module to have free access to the discussion area, thereby increasing the opportunity for collaborative discussions between different departments. The space atmosphere created in the cultural bearing mode is used to generate the atmosphere, and the cultural bearing mode of module design is used to promote the communication and cooperation between people. (Figure3-18)

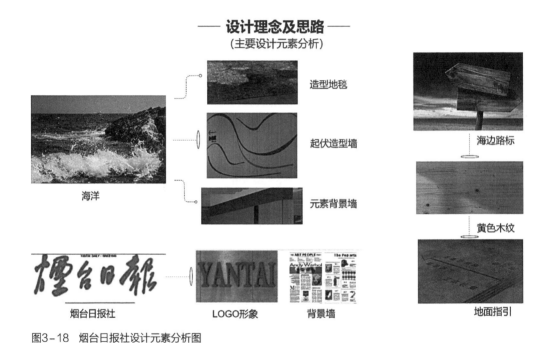

图3-18　烟台日报社设计元素分析图

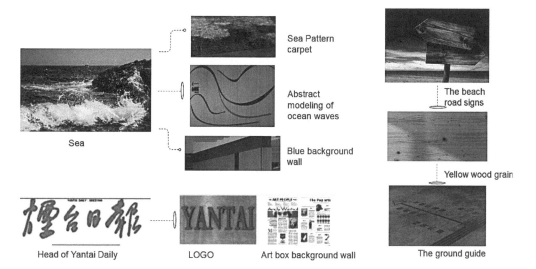

Figure 3 – 18 Analysis Diagram of Design Elements of the Daily Newspaper Office

整个办公空间以代表海洋的深蓝色为背景，以代表沙滩的浅黄木色作为家居和墙壁的部分饰面，通过颜色来表现烟台海洋城市的地理风貌（图3-19）。

图3-19 日报社项目文化模块南北向效果图
Figure 3-19 Renderings of the Cultural Module from South to North

The whole office space takes the dark blue color representing the ocean as the background, and the light yellow wood color representing the beach as part of the decoration of the home and walls. These colors are used to express the geographical features of the Yantai Ocean City. (Figure 3 – 19)

除了文化模块之外，空间中还设计了
两个休闲区和一个洽谈区，为员工和客户提
供更多的交流场所，迎合日报社的企业文化
（图3-20~图3-23）。

图3-20　前台
Figure 3-20　The Reception Desk

In addition to the cultural module, two leisure areas and a meeting area are designed in the space to provide more areas for staff and customers to communicate, catering to the Yantai Daily's corporate culture. (Figure 3–20~Figure 3–23)

图3-21 休闲区1
Figure 3–21 Leisure Area 1

图3-22 休闲区2
Figure 3-22 Leisure Area 2

图3-23 洽谈区
Figure 3-23 Discussion Area

艺术模块将烟台日报社25年来发行的各种纸质出版物在墙面上进行布置粘贴，使艺术盒子充满日报社的企业文化元素。艺术模块内有一个小型活动区，为员工提供文化、娱乐活动场所（图3－24）。

The art module is pasted with all kinds of paper publications issued by the Yantai Daily in the past 25 years, making the art module full of corporate cultural elements of the Daily. There is a small activity area in the art module, which provides a place for recreational activities for employees. (Figure 3–24)

图3-24 艺术模块入口效果图
Figure 3-24 Art module entrance

在文化模块的西墙外侧设计了抽象的海浪起伏造型，蓝色和浅黄色的搭配呼应了其他室内空间，迎合了设计主题（图3-25）。综上所述，作为文化载体，文化模块使用了造型设计和空间氛围营造的文化承载方式，集企业文化元素和海洋地理文化为一体，成为支撑整个空间设计主题的模块化建筑体。

On the outside of the west wall of the culture module, an abstract wave – undulating shape is designed. The blue and light yellow collocation echoes the other interior spaces and caters to the design theme(Figure 3 – 25). To sum up, as a cultural carrier, the cultural module adopts the cultural carrying mode of modeling design and space atmosphere creation, which integrates corporate cultural elements and marine geographical culture, and thus become a modular building body supporting the whole space design theme.

图3-25 烟台日报社项目文化模块效果图
Figure 3 – 25 Renderings of the Cultural Module of the Yantai Daily Design Project

3.3　本章小结

　　模块作为文化的载体，在四个项目中的文化承载方式各不相同，充分展现了模块建筑设计法的灵活性、多样性等特点。在表现形式上，模块化建筑也拥有更丰富的承载方式，从具象到抽象来体现本土文化内涵。模块化建筑设计法的灵活性和多样性还体现在单一文化承载方式的不同表现上。例如在商业书店项目中，同样使用造型设计的文化承载方式，也会因为设计师的设计想法和客户诉求而呈现出不同的空间效果。因此，在灵活、多样的文化承载方式下，模块化建筑设计法能够更好地适应客户和设计师的需求，脱离传统的僵化工业建筑模块，在本土文化的继承和发展上发挥自己的作用。

3.3 Conclusion

As the carrier of culture, modules are used in different ways to carry culture in the four design projects, which fully demonstrates the flexibility and diversity of module architecture design method. In the expression form of cultural bearing, modular architecture also has a richer way of bearing, from concrete to abstract to reflect the connotation of local culture. The flexibility and diversity of modular architectural design method is also reflected in the different performance of single culture bearing mode. For example, compared with the commercial bookstore project, the cultural bearing mode of modeling design is also used, which presents different spatial effects due to the designers' ideas and customer demands. Therefore, under the flexible and diverse cultural carrying modes, modular architectural design method can better adapt to the design needs of customers and designers. This design method can break away from the traditional rigid industrial building modules, and play its role in the inheritance and development of local culture.

4 模块化建筑设计法的创新应用

4.1 现阶段模块化设计方法的应用概况

在装配式住宅的设计研究中，根据功能、结构、大小等不同特点设计了三类模块——基本功能模块、核心筒模块、配套模块。各模块在设计时应考虑功能性、舒适性、通用性、规则性等，同系列之间具有一定的逻辑性和衍生性，便于多样化的组合，形成满足不同需求的住宅产品（任钟颖，2021）。

4.1.1 基本功能模块

基本功能模块是住宅的主要组成部分，包含客厅、餐厅、卧室、厨房、卫生间等多个功能空间，需要设计出不同规格尺寸的功能模块。这些功能模块应尽量采用相同开间或进深尺寸，既相互独立又能快速组合，具有一定的逻辑性和衍生性，便于组合成多样化的单元模块。（任钟颖，2021）

单元模块是平面设计中最重要的组成部分，通过设计集成灵活布置基本功能模块。结合不同家庭结构和人群的使用需求，基本功能模块可自由组合成两室、三室、四室的单元模块，再进一步组合变换形成多样化的户型空间，结合实际情况对应满足不同居住建筑的户型和面积要求。（任钟颖，2021）

4.1.2 核心筒模块

核心筒模块是平面设计中的重要组成部分，不仅要考虑其经济性和舒适性，合理地布局电梯、候梯厅、水电管井、公共廊道等空间，还要考虑其是否有利于装配式结构布置以及如何与其他模块进行组合。因此，设计的核心筒模块形状规整、结构对称，便于组合衍生。（任钟颖，2021）

4 Innovative Applications of Modular Architectural Design Methods

4.1 The Application of Modular Design Method at Present

In the research of prefabricated housing, three types of modules are designed according to the different characteristics of function, structure and size — basic function module, core tube module and supporting module. Each module design should consider the factor of functionality, comfort, universality, regularity, etc., with a certain logic and derivative between the series, easy to diversify the combination, the formation of residential products to meet different needs (Ren Zhongying, 2021).

4.1.1 Basic Functional modules

The basic function module is the main component of the house, including living room, dining room, bedroom, kitchen, toilet and other functional space, which need different specific design of the size of the function module. These functional modules need to adopt the same size of opening or depth as far as possible. They are independent and can be combined quickly, with certain logic and derivation, which is easy to combine into diversified unit modules.

Unit module is the most important part of graphic design. Basic functional modules can be arranged flexibly through design integration. Combined with the use needs of different family structures and people, basic functional modules can be freely combined into two–room, three–room and four–room modules, and then further combined and transformed to form diversified apartment space, which can meet the requirements of different residential buildings in terms of apartment type and combined the area with actual situation (Ren Zhongying, 2021).

4.1.2 Core tube modules

Core tube module is an important part of the graphic design, We not only need to consider its economy and comfort, reasonable layout of building elevator, waiting hall, water and electricity pipe well, public corridor and other spaces, but also need to consider whether it is conducive to assembly structure layout and how to combine it with other modules. Therefore, the design of the core cylinder module shape needs to be regular, symmetrical structure, and easy for combination and derivation (Ren Zhongying, 2021).

4.1.3 配套模块

除了基本功能模块和核心筒模块，还有配套模块，作为单元模块和核心筒模块共同的补充模块，可用作生活阳台或玄关等，尺寸规格尽量统一。

4.1.4 现阶段模块化设计法的优势

通过上述模块化设计的主要组成模块类型可以看出，目前使用的模块化设计法依然保有传统装配式住宅设计的标准化、工业化、参数化等特点。同时，模块化设计作为传统装配式建筑设计方式的进化产物，能够在传统装配式建造中的基础上推陈出新，具备更多的灵活性。从设计的角度来说，模块化简化了设计，增强了设计的通用性和灵活性，提升了设计效率。从生产运输施工来讲，模块化设计大大减少了构件规格种类，规整了构件尺寸，使得生产运输更为高效，施工更为便捷。

4.1.3 Supporting modules

In addition to the basic function module and core tube module, there is also the supporting module, as a complementary module of the unit module and core tube module, which can be used as a living balcony or porch, the size and specifications are as unified as possible.

4.1.4 Advantages of Modular Design Method at Present

Through the above modular design modules, it can be analyzed that, the modular design method used at present still retains the characteristics of standardization, industrialization and parameterization of traditional prefabricated housing design. At the same time, modular design, as the evolution of the traditional assembly building design, can bring forth the new on the basis of the traditional assembly building with more flexibility. From the view of design, modularization simplifies the design, enhances the versatility and flexibility, and improves the design efficiency. In terms of production and transportation, the modular design greatly reduces the types of component specifications, regularizes the size of components, and makes production and transportation more efficient and the construction more convenient.

4.2 模块化设计法的创新应用

受限于目前的技术手段，在技术上对现阶段的模块化设计法进行创新颇为困难。但是，从模块的空间应用角度考虑，仍然有创新的空间。本书提出的模块化建筑设计方法继承了模块化建筑的灵活性，在建筑空间和室内空间等不同空间尺度下都可灵活使用。同时，模块个体也可以根据实际空间尺寸自由控制模块体积的大小。不同于传统模块化建筑，模块化建筑设计法在项目中的应用创新点主要有以下几点：

（1）空间改造创新应用：室内空间尺度下，可以利用模块置入，对室内空间进行分层处理，把平层变为多层空间，提高空间利用率。

（2）空间分隔创新应用：在写字楼等大平层空间，可以利用半开敞模块空间的置入，自由分隔功能区，减少室内墙体对室内空间通风采光的阻隔，使空间更加开敞通透，简化施工工序，分区合理化。

（3）模块墙体材料创新应用：随着工程技术水平的提高，越来越多的新型承重材料被使用，玻璃、秸秆等轻型材料也可以满足承重需求。同时，适用材料的种类增多也使多样化的模块造型成为可能，这些都为设计师的设计提供了更多的可实施性。

（4）结合建筑节能技术的应用：绿色建筑一直是国际建筑研究的主题之一。绿色节能技术也在不断进步。模块化建筑设计法与绿色建筑的结合能够更好地应用于乡村改造建设当中，减轻当地居民的经济负担和大规模建筑作业造成的环境污染。

4.2 Innovative Application of Modular Design Method

Due to the limitations of the current technical means, the innovation of the modular design method is very limited. However, there is still room for innovation in the spatial application of modules. The modular building design method proposed in this research inherits the flexibility of modular building and can be used flexibly in different spatial scales such as architectural space and interior space. At the same time, the size of the module volume of individual module can be freely controlled according to the actual space size. Different from the traditional modular building, the application innovation of modular building design method in the project mainly includes the following points:

(1) Innovative application of space transformation: at the scale of interior space, modules can be placed according to the layered processing of interior space, transforming flat layer into multi-layer space and improving space utilization rate.

(2) Innovative application of space separation: in large flat floor space such as office buildings, semi-open module space can be used to separate functional areas freely, reduce the indoor wall barrier for ventilation and lighting, make the space more open and transparent, simplify the construction process, and rationalize the partition.

(3) Innovative application of modular wall materials: with the improvement of engineering technology, more and more new load-bearing materials are used, and light materials such as glass and straw can also meet the load-bearing requirements. At the same time, the variety of applicable materials also makes it possible to diversify the modular modeling, which provides more possibilities for the designer's ideas.

(4) Combined application of building energy saving technology: Green building has always been one of the themes of international architectural research. Green energy saving technology is also constantly improving. The combination of modular building design method and green building can be better applied in rural reconstruction to reduce the economic burden of local residents and the environmental pollution caused by large scale construction operations.

4.3 方案分析

4.3.1 北方印宅建筑设计项目

北方印宅建筑设计项目是纯模块化搭建的个人住宅。该项目以研究探索为目的，对不同的功能区进行组合搭建，是模块化建筑设计方法在建筑尺度层面的案例。北方印宅项目充分发挥了模块化建筑设计方法的灵活性，可以根据乡村改造的实际情况按需进行功能区的删减。北方印宅建筑设计项目是对模块化建筑在建筑大尺度空间创新应用和结合绿色节能技术的探索研究。通过该项目的实际建造可以发现，模块化建筑的空间分隔创新应用和模块墙体材料创新应用在建筑大尺度空间中依然可行。北方印宅项目使用的秸秆板材、太阳能光伏建筑一体化等技术，都是模块化建筑与绿色节能技术结合应用的方式。两者的成功结合为模块化建筑的普及与发展提供了良好的技术支持。

通过北方印宅的模块搭建工序，可以看出模块的组成很灵活，可以根据设计造型和具体的使用需要进行模块制造。模块的大小、数量都可以调整，这也为设计师想法的实施提供了必要的技术基础。同时，该项目将应用于中国乡村改造，基于每户家庭的成员人数、经济条件等情况各不相同，模块化建筑可以充分发挥其灵活性，相应进行增减，来满足个体住户对功能区的不同使用需求。（图4–1）

4.3 Design Projects Analysis

4.3.1 Yin House Architectural Design Project

Yin House architectural design project is a pure modular building personal residence. For the purpose of research, the project combines different functional areas. It is a case research of modular architectural design methods at the architectural scale level. The Yin House project gives full play to the flexibility of modular architectural design method, and can be reduced according to the actual situation of rural reconstruction. Yin House architectural design project is an exploration on the innovative application of modular building in large scale space and the combination of green energy saving technology. Through the actual construction of the Yin House project, it can be found that the innovative application of modular building space separation and modular wall material is still feasible in the large scale space of the building. The technologies used in the Yin House project, such as straw board and solar photo-voltaic building integration, are the combined application methods of modular building and green energy-saving technology. The successful combination of the two provides a technical support for the popularization and development of modular architecture.

Through the module building process of the Yin House, it can be told that the composition of the module is flexible, and the module can be manufactured according to the design and modeling and the specific use needs. The size and number of modules can be adjusted, which provides the necessary technical basis for the implementation of designers' ideas. At the same time, the project will be applied to the rural reconstruction of China. Based on the number of members of each family, economic conditions and other different conditions, modular building can give full play to its flexible characteristics. Through the corresponding increase or decrease of the number of modules, to meet different needs of individual households for the functional area. (Figure 4-1)

图4-1 北方印宅的模块搭建工序图
Figure 4-1 Yin House Module Construction Process

绿色建材——秸秆板材

绿色建材是指健康、环保、安全的建材，在世界范围内又被称为"健康建材"或"环保建材"。北方印宅模块的承重墙体主要以秸秆板材为主，秸秆材料的使用是对模块墙体材料创新应用的典型例证。印宅建筑项目使用的秸秆板材属于环保建材。秸秆板材是以秸秆等农作物的废弃物为原料，生产过程无须进行废弃物排放。建筑材料具有重量轻、强度高、承重性强等特点，是一种环保性能高的新型绿色建筑材料。（图4-2）

图4-2　北方印宅的秸秆板材和墙面绿化
Figure 4-2　Yin House Straw Board and Wall Afforestation

太阳能与光伏建筑的融合是太阳能发电的一种新概念。简单地说，太阳能光伏组件阵列安装在建筑的外表面获取太阳能进而转化为电力，为建筑提供能源。太阳能与光伏建筑的融合产生绿色能源，应用于太阳能发电，不会污染环境。光伏阵列一般安装在闲置的屋顶或外墙上，不占用额外的土地，这对于昂贵的城市建筑来说尤为重要。夏季是用电高峰季节，也是日照量最大的时期，因此光伏系统发电量能够达到全年最大峰值，在电网上起到调峰作用。而太阳能光伏建筑集成技术采用光伏并网系统，不需要电池，节省投资，不受电池充电状态的限制，可以充分利用光伏系统产生的能量。光伏阵列组件吸收太阳能并将其转化为电能，降低了室外综合温度，降低了墙体的热量增益和室内空调的冷负荷，因此也可以起到建筑节能的作用。（图4-3）

图4-3　印宅效果图：太阳能光伏建筑一体化
Figure 4-3　Perspective Rendering of the Yin House: Solar Photo-Voltaic Building Integration

Green Building Materials – Straw Board

Green building materials refer to healthy, environmentally – friendly and safe building materials. It is also known as "healthy building materials" or "environmental building materials". The load – bearing wall of the Yin House module is mainly made of straw board, and the use of straw material is a typical example of the innovative application of the module wall material. The straw board used in this building is environmental protection building materials. The straw board uses straw and other crop waste materials as raw materials without the process of wastes discharge and production.The building materials,which have light weight, high strength and good load – bearing performance, are new green building materials with high environmental protection performance. (Figure 4 – 2)

Integration of solar energy and photo – voltaic architecture is a new concept of solar power generation. To put it simply, the solar photo – voltaic array is installed on the outer surface of the envelope of the building to provide electricity. Integration of solar energy and photo – voltaic architecture produces green energy, which is applied to solar power and will not pollute the environment. It is also a renewable energy source that is inexhaustible. Photo – voltaic arrays are generally installed on idle roofs or external walls without additional land occupation. This is particularly important for expensive urban buildings. Summer is the peak season for electricity use and the period with the maximum amount of sunlight, thus photo – voltaic system generate the most electricity and play a peaking role on the power grid. And solar photo – voltaic building integrated technology adopts the grid connected photo – voltaic system, which do not need for a battery, and could saves investment without limited by the state of charge of the battery. It can make full use of the power generated by the photo – voltaic system. The photo – voltaic array absorbs solar energy and converts it into electricity, which reduces the outdoor comprehensive temperature, reduces the heat gain of the wall and the cooling load of the indoor air conditioning, and therefore can also play a role in building energy conservation. (Figure 4 – 3)

4.3.2 朝阳街博物馆设计方案

朝阳街博物馆设计项目主要使用了模块化建筑设计法中的空间改造应用和空间分隔应用。模块的置入将两层的博物馆建筑体分隔为不同大小的展厅。而独立展厅根据不同的历史内容展示需要，对墙体进行半开敞处理，既保证了展厅与展厅之间交通动线的流畅，又起到了分隔展厅的作用。（图4-4）

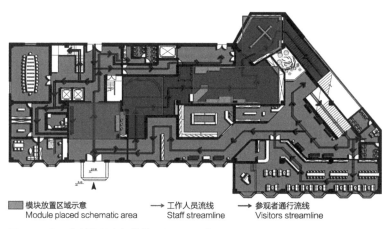

■ 模块放置区域示意 　　→ 工作人员流线 　　→ 参观者通行流线
Module placed schematic area 　Staff streamline 　Visitors streamline

图4-4 朝阳街博物馆空间模块置入平面示意图
Figure 4-4 Module Insert Schematic Diagram of Chaoyang Street Museum

朝阳街开埠文化始于1961年侵略者乘船抵达烟台港口。因此序厅模块主要以清水混凝土搭建的船帆造型作为墙体，营造出乘风破浪的氛围来阐述开埠文化的开端。这个模块的文化内容较少，所以模块空间尺度相比后面的展厅较小，于是为了展示空间的连贯性，使用了半开敞的模块构件。（图4-5）

如图4-6所示，开埠之初展厅主要展现了鸦片战争时期，外国侵略者通过战争手段打开中国港口的过程。此展厅用三个小模块相叠加组合成展厅之间的通道节点。每个小模块中置入枪炮等武器作为隐喻，展厅的入口地面用木板作为饰面模拟码头的造型。展厅的三段不规则通道设计展现出了模块化建筑设计方法的空间分隔创新应用。

4.3.2 Design Scheme of Chaoyang Street Museum

This design project mainly uses the modular building design method in the space transformation application and space separation application. The placement of modules divides the two-storey museum building into galleries of different space size. The independent exhibition hall makes the wall semi-open according to different needs of historical content display, which not only ensures the smooth traffic flow between each exhibition halls, but also plays the role of separating the exhibition hall. (Figure 4-4)

The culture of opening ports in Chaoyang Street began in 1961 when invaders arrived at Yantai port. Therefore, the introduction exhibition hall module mainly uses the sail shape built by bare concrete wall, creating an atmosphere of braving wind and waves to illustrate the port opening culture. The cultural content of this module is less, so the spatial scale of the module is smaller than other exhibition halls behind. In order to show the coherence of the space, semi-open module components are used. (Figure 4-5)

The preface hall mainly shows the process of the invaders opening China's ports through the Opium War. This exhibition hall is composed of three small modules which are superimposed to form the channel node between the exhibition halls. Guns and other weapons are placed in each small module as metaphors. The entrance floor of the preface hall is decorated with wooden boards to simulate the shape of a wharf. The design of three irregular passages in the preface hall demonstrates the innovative application of modular architectural design method to space separation. (Figure 4-6)

图4-5 序厅
Figure 4-5 The Introduction Exhibition Hall

图4-6 开埠之初展厅
Figure 4-6 Preface Exhibition

随着港口的打开，基督教文化也随着侵略者而传入中国，基督文化展厅介绍的就是基督教文化对朝阳街的影响。为了满足设计中光影十字架的呈现，该展区模块的空间尺度远大于前几个展厅。该模块的墙体围合并没有按照传统的长方体模块结构，而是根据设计方案的特殊造型进行围合，这也体现了模块化建筑在室内空间的个体独立性和灵活性。（图4–7）

咖啡厅的设计承接了之前基督教文化展厅的主要元素，在狭长的过道空间附加了顶棚的彩色玻璃饰面和清水混凝土墙面，将咖啡厅围合。模块化建筑墙体材料的多样性为设计师提供了更多的文化表现手法，让文化元素与建筑设计的融合不拘泥于几种固定的表现形式，避免设计思维僵化。（图4–8）

图4–7　基督教文化展厅
Figure 4–7　Christian Introduction Exhibition

图4-8 咖啡厅
Figure 4-8 Cafe

With the opening of the harbor, Christian culture was introduced into Yantai along with the invaders. The Christian exhibition hall introduces the influence of Christian culture on Chaoyang Street. In order to satisfy the presentation of the cross of light in design, the spatial scale of the exhibition area module is much larger than those of the previous exhibition halls. The wall enclosure of this module is not in accordance with the traditional cuboid module structure, but according to the special shape of the design scheme for enclosure, which also reflects the modular building in the interior space of individual independence and flexibility. (Figure 4-7)

The design of the Cafe carries on the main elements of the previous Christian culture exhibition hall. The long and narrow corridor space is enclosed by the additional stained glass for the ceiling and bare concrete wall. The diversity of modular building wall materials gives designers more cultural expression techniques, so that the integration of cultural elements and architectural design is not restricted to a few fixed forms of expression, and so that to avoid the rigidity of design thinking. (Figure 4-8)

4.3.3 商业书店设计项目

商业书店设计项目是模块化建筑设计方法在小尺度室内空间应用的典型案例。利用模块化建筑自身的承重能力和自由组合的灵活性，将一层的室内空间分割为两层，增加了空间的利用率，同时对每个功能区的文化主题规划进行了整理。基于客户的要求和书店的商业属性，商业书店中要增加科幻读物阅读区的主题设计，但是整个设计的主题还要以中国传统文化为主。面对古代元素与现代元素的格格不入，模块化建筑设计方法的独立性优势就显现出来。既满足了不同功能区的文化特色，又保持了整体的文化韵味。让难以融合的各个区域能够服务于整体设计，保持设计主题的一致性。

图4-9～图4-11为商业书店的一层平面图与空间模块置入示意图。根据一层平面图可以看出，一层靠北处黑色部分为增加的一个长方形模块空间，对空间进行了抬升，在此基础上形成了第二层结构。靠近中央的棕色部分为阶梯式书架，以斜坡式模块进行一层与二层的衔接。

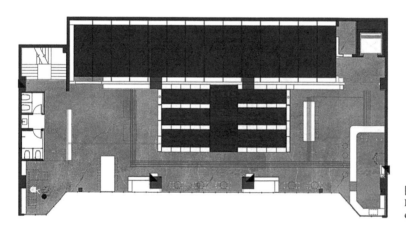

图4-9 书店一层平面图
Figure 4-9 First Floor Plan of Bookstore

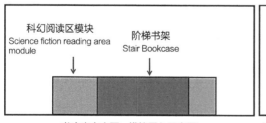

书店南向立面：模块置入示意图
South Facade of Bookstore

图4-10 书店空间模块置入示意图
Figure 4-10 Schematic Diagram of Bookstore Space Module Placement

图4-11 科幻主题阅读区
Figure 4-11 Arch-Shaped Book Shelves

4.3.3 Commercial Bookstore Design Project

The Commercial bookstore design project is a typical case of modular architectural design method applied in small scale interior space. Using the modular building's own load-bearing capacity and the flexibility of free combination, the interior space on the first floor is divided into two levels, increasing the utilization of space, and at the same time, the cultural theme planning of each functional area is organized. Based on the requirements of customers and the commercial properties of the bookstore, the reading area of science fiction should be added in the commercial bookstore, but the theme of the whole design should be based on the traditional Chinese culture. Faced with the incompatibility between traditional and modern elements, the independent advantage of modular architectural design method appears. It not only meets the cultural characteristics of different functional reading areas, but also maintains the overall cultural atmosphere. Modular architectural design method makes each different design theme area of the bookstore to serve the overall design and maintain the consistency of the design theme.

The following figure shows the first floor plan and spatial module placement sketch of the commercial bookstore. According to the plan of the first floor, it can be seen that the black part near the north of the first floor is an additional rectangular module space, which is lifted, and on this basis, the second floor structure is formed. The brown part near the center is a stepped bookcase, connecting the first and second floors with sloping modules. (Figure 4-9~Figure 4-11)

　　根据客户的要求，在以中国传统文化为主题的阅读空间中加入科幻元素阅读区。为避免设计主题混乱，将科幻主题阅读区设计为一个单独模块，入口设置在书店门口附近。整个科幻阅读区的结构设计为拱形，中间创造出裂痕，置入筒灯进行照明。地面使用黑色理石，通过映射顶棚光源，营造宇宙的空间感。因此，模块化建筑设计方法不仅适用于传统文化承载，对于其他的文化元素同样适用。科幻阅读区的模块置入展现出了模块化建筑设计方法的空间改造创新应用。

　　书架阶梯是对科幻主题模块空间的补充式设计，在有限的空间里减少二层台阶的占地面积，尽可能增加室内空间的使用面积和书籍的存储空间。同时书架阶梯的斜坡形式遮掩了科幻模块的外部结构，使模块的空间置入不突兀。（图4–12）

　　这种模块与模块之间的组合搭配也是本研究对模块化建筑设计方法灵活应用的一种方式。模块化建筑设计法同样可以应用于小型尺度空间的造型设计当中，且能对空间结构进行改进，提高空间利用率。当遇到需要融合不同文化主题的情况，也可以保持设计主题的整体性。

图4–12　阶梯书架
Figure 4–12 Stair Book Shelves Perspective

According to the requirements of customers, science fiction reading area is added to the reading space. To avoid the confusion of the design theme, the science fiction reading area is designed as a separate module, and the entrance is set near the entrance of the bookstore. The structure of the entire science fiction reading area is arched, with cracks created in the middle, and lit by downlights. The ground is made of black marble, which reflects the ceiling light source to create a sense of space in the universe. Therefore, modular architectural design method is not only applicable to traditional culture, but also applicable to other cultural elements. The placement of modules in the science–fiction reading area demonstrates the innovative application of modular architectural design methods to the spatial transformation.

The bookshelf ladder is a supplementary design to the space of the science–fiction theme module. In the limited space, the occupation area of the second floor step is reduced, and the use area of the interior space and the storage space of the books are increased as much as possible. At the same time, the sloping form of the bookshelf steps hides the external structure of the science fiction module, so that the space of the module is not obstructive. (Figure 4–12)

This combination of modules is a flexible way of using modular architectural design method. Modular architectural design method can also be applied to the modeling design of small scale space, and can improve the spatial structure and the utilization rate of the space. When different cultural themes need to be integrated, the integrity of the design theme can also be maintained.

4.3.4 烟台日报社办公空间设计项目

烟台日报社办公空间设计项目主要是对企业文化元素的继承与融合，同时在这个设计项目中还融合有烟台地理位置的元素特征，比如海浪、集装箱等元素。根据客户的要求，设计师在空间中置入了三个不同大小的模块：艺术模块（Art Module）、生态模块（Ecological Module）、文化模块(Cultural Module)。三个模块分别以玻璃全景和局部开敞的方式让三个模块与整个室内空间能够进行沟通，达到在空间中的任何一个位置都能看到至少一个模块的效果。不同于商业书店设计项目的独立封闭模块，烟台日报社设计项目的三个主题模块被设计得极其开敞，增强模块与周边空间的交流，以期让更多的员工能够享受到舒适、绿色的办公环境和开阔的视野。（图4-13）

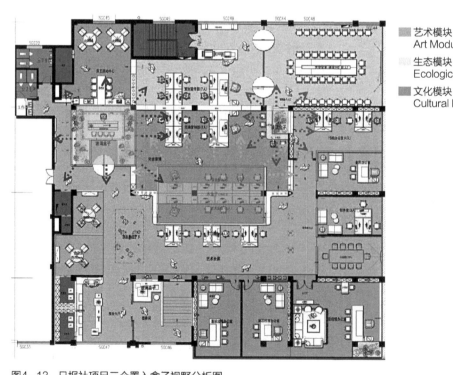

■ 艺术模块
Art Module

生态模块
Ecological Module

■ 文化模块
Cultural Module

图4-13 日报社项目三个置入盒子视野分析图
Figure 4-13 Visual field analysis diagram of three modules placed in Yantai daily design project

4.3.4 Yantai Daily Office Design Project

The Yantai Daily office design project is mainly the inheritance and integration of corporate culture elements. At the same time, the design project integrates geographical elements of Yantai, such as waves, containers and other elements into the design. According to the client's requirements, the designer put three modules of different sizes into the space: Art Module, Ecological Module and Cultural Module. The three modules are connected with the whole interior space through glass panorama, and each module is partial open, so that at least one module can be seen in any position in the office space. Different from the independent closed module of the commercial bookstore design project, the three theme modules of the Yantai Daily office design project are designed to be extremely open, to enhance the communication between the module and the surrounding space, in order to allow more employees to enjoy a comfortable, green office environment and a wide vision. (Figure 4 – 13)

　　文化模块居于整个办公空间的中心位置，是员工集中讨论交流的功能区，也是企业文化所注重的部分，所以被称为文化模块。考虑到通风采光还有员工沟通交流等问题，文化模块的构造变为南北通透的造型。将讨论区置入其中，有助于员工之间的工作交流和不同部门的协作，提高工作效率。同时文化模块的置入也对中央大厅的所有办公区域进行了划分，使本来毫无规律可循的空间被划分为四个区域，替代了传统的墙壁分区方式，这也是本项目对传统模块化建筑设计法的创新。在作为文化载体的职能方面，模块的外表面使用了抽象的海浪图案，使用功能上也重点突出了企业文化中的团结合作精神。文化模块的置入展现出了模块化建筑设计方法的空间分隔创新应用。（图4-14）

图4-14　日报社项目文化模块与办公区效果图
Figure 4-14　Rendering of the Cultural Module and Office Area of the Yantai Daily Design Project

The culture module is located in the center of the whole office space. It is the functional area where employees focus on discussion and communication. It is also the part that Yantai Daily company culture pays attention to, so it is called culture module. Considering the problems of ventilation, lighting and staff communication, the structure of the cultural module is changed into a north-south oriented no-walls shape. The discussion area is put into it, which helps the communication between employees and the collaboration between different departments, and improves the work efficiency. At the same time, the cultural module also divides all the office areas of the central hall, so that the previously irregular space is divided into four areas, replacing the traditional wall partition method, which is also the innovation of this project to the traditional modular architectural design method. As a cultural carrier, the outer surface of the module uses an abstract sea wave pattern, and the functions also highlights the spirit of unity and cooperation in the corporate culture. The placement of cultural modules shows the innovative application of spatial separation in modular architectural design methods. (Figure 4-14)

艺术模块使用了全玻璃门窗，保证室内空间通风采光充足。绿植的引入也可舒缓员工的工作压力。艺术模块的墙壁为三面玻璃、一面木质墙壁。艺术模块使用的玻璃墙壁材料是对模块墙体材料创新应用的体现。（图4－15）

Full glass doors and windows are used in the art module to ensure adequate ventilation and lighting in the interior space. The introduction of green plants can relieve staff's work stress. The walls of the art module are glass on three sides and wood on one side. The glass material used in the art module wall is the embodiment of the innovative application of the modular wall material. (Figure 4 – 15)

图4 – 15　艺术模块西向效果图
Figure 4 – 15　Art Module Westward Rendering

生态模块与艺术模块东西相对，使用玻璃幕墙，其功能为弥补空间的视野死角，让办公的员工能够最大限度地享受绿植的景色，舒缓精神压力。生态模块同样使用了玻璃壁材料，体现了对模块墙体材料的创新应用。（图4–16、图4–17）

图4–16　生态模块效果图
Figure 4–16 Ecological Module Rendering

图4–17　生态模块、艺术模块与办公区效果图
Figure 4–17 Ecological Module, Art Module and Office Area Rendering

The ecological module is opposite to the art module, and the glass curtain wall is used, as well. Its function is to make up the dead corner of the space, so that the office staff can enjoy the green scenery to the greatest extent and relieve the mental pressure. The ecological module also uses glass wall materials, reflecting the innovative application of modular wall materials. (Figure 4 – 16, Figure 4 – 17)

4.4　本章小结

　　模块化建筑设计方法在上述四个设计项目中展现的创新应用各不相同。不论是在北方印宅这种大尺度建筑空间，还是商业书店这种小尺度室内空间，都有其应用的创新点。在技术的支持下，模块造型和材料的使用种类都更加多元化，改善了传统模块化建筑的千篇一律，能够更好地满足设计师的设计想法，设计出更有设计感和文化内涵的建筑作品。

　　模块化建筑设计是建筑设计施工行业的产业升级转型，在短短五六年的时间里，中国已有大量项目采用装配式建造方式。上述项目的实践证明，模块化建筑设计法具有高效、质优、便宜、绿色低碳等优点，除了能很好地适应社会快速发展的需要之外，对环境的影响也较传统建造方式更友好，呈现出极好的发展趋势。本书从设计师角度出发，思考模块化建筑设计法的创新应用，并提出使用策略，有助于设计师挖掘模块化设计法的应用潜力，帮助设计师用自己的专业能力，迎接模块化建筑设计的发展浪潮，并推动模块化建筑设计在当代的进一步发展。（李燕红，2021）

4.4 Conclusion

The innovative application of modular architectural design methods in the above four design projects is different. No matter in the large scale architectural space such as Yin House, or in the small scale indoor space such as commercial bookstore, all of these design projects have innovation points in their application. With the support of technology, the types of modules and materials used are more diversified than before, which improves the uniformity of traditional modular buildings, and can better meet the design ideas of designers to make architectural design works with more sense of design and cultural connotation.

Modular architectural design is the industrial upgrading and transformation of architectural design industry. In just five or six years, a large number of projects in China have adopted the prefabricated construction method. The practice of the above projects has proved that the modular building design method has the advantages of high efficiency, high quality, low cost, green and low carbon. In addition to being well adapted to the rapid development of society, it also has a more friendly impact on the environment than the traditional construction method, showing an excellent development trend. This paper, from the designers point of view, thinks about the innovative application of modular architectural design method, and puts forward the use strategy, which will help designers to tap the application potential of modular design method, help designers to use their professional ability, meet the development tide of modular architectural design, and promote the further development of modular architectural design in the contemporary world (Li Yanhong, 2021).

5 模块化建筑空间的连接方式

现阶段国内外普遍使用的建筑模块之间的连接方式包括焊接、连接板或连接件螺栓连接。但是传统的模块连接策略都是针对模块之间的物理连接手段，在空间尺度和视野连接方面并没有太多相关研究。

模块化建筑设计方法除了在应用和文化承载方式上具有灵活性和多样性，在建筑空间、室内空间等不同空间尺度上模块的连接方式也具有灵活性和多样性。为了满足不同设计要求和客户要求，单一的模块连接方式已不足以满足复杂的空间组合变化。因此，根据模块化建筑设计方法的特点，本书提出的主要连接方法包括交通流线与视点的连接、模块与模块的连接、模块与不同尺度空间的连接。其中，模块与模块的连接方式遵循传统模块化建筑的连接策略，在这些研究项目的实践过程中总结了交通流线与视点的连接方式以及不同尺度的模块与空间的连接方式，这是传统模块化建筑所没有的。

5.1 模块与模块的连接方式

模块与模块的连接方式在模块化建筑的理论综述部分有所涉及。本连接方式以新型模块化建筑的组装方式作为其连接策略。该连接方式主要是解决模块与模块之间大尺度建筑空间的衔接与支撑问题。在轻型墙体材料的技术革新背景下，建筑的建造过程如同搭积木，每个模块之间利用焊接、铆钉等方式进行衔接即可。施工过程简便，运输成本低，能够根据设计师的想法灵活搭配。工业化生产固定模块结构，不需要针对每个项目重新制作模具，降低了生产成本。北方印宅建筑设计项目就是使用模块与模块连接的典型案例。

5 The Connections of Modular Architectural Spaces

At present, the commonly used connection modes between building modules includes welding, connecting plate or connecting piece bolted connection. However, the traditional module connection strategy is aimed at the physical connection between modules, and there is not much research on spatial scale and visual field connection.

In addition to the flexibility and diversity in the application and cultural bearing modes, modular architectural design method also has flexibility and diversity in the connection modes of modules in different spatial scales such as architectural space and interior space. In order to satisfy different design requirements and customer requirements, a single module connection mode is not enough to meet the complex spatial combination changes. Therefore, according to the characteristics of modular building design method, the main connection methods proposed in this book include: the connection between traffic flow line and viewpoint, the connection between modules, and the connection between modules and space at different scales. Among them, the connection mode of between modules and follows the connection strategy of traditional modular buildings, and the connection mode of traffic flow and viewpoint and the connection mode of modules and space at different scales are summarized in the practice process of these research projects, which are the connection modes that traditional modular buildings do not have.

5.1 Connection Mode Between Modules

The connection mode between modules is discussed in the literature review of modular architecture in chapter 1. The connection method takes the assembly method of the new modular building as its connection strategy. This connection mode is mainly to solve the problem of connecting and supporting the large scale building space between modules. In the context of technological innovation of lightweight wall materials, the construction process of the building is like building blocks. Each module is connected by means of welding and riveting. The construction process is simple, the transportation cost is low, and it can be flexibly matched according to the designer's idea. Industrial production of fixed module structure do not need to re-make the mold for each project, and this can reduce the production cost. Yin House architectural design project is a typical case of using the connection mode between modules.

模块与模块的连接方式是基于建筑空间的尺度来选取的，根据新型模块化建筑的装配形式，该连接方式的实际施工模式细化为以下几种：

（1）全盒式，完全由承重盒子重叠组成建筑。

（2）板材盒式，将小开间的厨房、卫生间或楼梯间等做成承重盒子，再与墙板和楼板等组成建筑。

（3）核心体盒式，以承重的卫生间盒子作为核心体，四周再用楼板、墙板或骨架组成建筑。

（4）骨架盒式，用轻质材料制成的许多住宅单元或单间式盒子，支承在承重骨架上形成建筑。也有用轻质材料制成包括设备和管道的卫生间盒子，安置在用其他结构形式的建筑内。

北方印宅建筑设计方案选用的是全盒式装配形式。在工厂制造阶段，用钢结构框架固定秸秆板材墙体，在连接处提前预留铆钉固定点，构建每个独立模块。在模块运输到施工现场后，工人用铆钉在预留的位置上进行安装，并用焊接的方式将钢结构进行固定，最终完成模块与模块之间的连接。（图5–1）

图5–1　北方印宅全盒式装配搭建工序图
Figure 5–1 Yin House Full Box Assembly Construction Process

The connection mode of modules is selected based on the scale of the building space. According to the assembly form of the new modular building, the actual construction mode of the connection mode is refined into the followings:

(1) Full box type, completely composed by load-bearing boxes overlapping together to form the building.

(2) Board box type, the small kitchen, bathroom or stairwell are made into load-bearing boxes, and then add wall panels and floors and other components to complete the whole building.

(3) The core body box type, with the load-bearing toilet box as the core body, surrounded by the floor, wall or skeleton composition to form the whole building.

(4) Skeleton box, a number of residential units or single-room boxes made of lightweight materials, supported by a load-bearing skeleton to form a building. Another way is to use lightweight materials to make toilet boxes containing equipment and plumbing and then placed them in buildings using other structural forms.

The Yin House adopted the full box assembly mode. In the factory manufacturing stage, steel frame is used to fix the straw board wall, and rivet fixing points are reserved in advance at the joints to build each independent module. After the modules are transported to the construction site, the workers install them in the reserved position with rivets, and fix the steel structure through welding. Then the connection between the modules is completed. (Figure 5-1)

5.2 模块与空间的不同尺度连接方式

　　模块与空间的不同尺度连接主要适用于室内空间的空间重塑。这里所说的空间重塑，指的是利用模块的可控边界对现有空间进行重新塑造，以达到符合设计要求的空间组成。这种连接方式适用于提高空间利用率、划分功能区、塑造特定空间造型等设计需求。商业书店项目和烟台日报社办公空间设计项目都使用了这种连接方式。

5.2.1 商业书店设计项目

　　由图5-2和图5-3可以看出，商业书店项目在一层靠北处增加了一个黑色长方形模块，对整层空间进行了重新塑造，把原本一层的室内空间扩增至两层。这种根据室内空间实际情况进行空间重塑的方式，是模块与不同空间尺度连接方式的实际应用案例，展现了该连接方式提高空间利用率的作用。

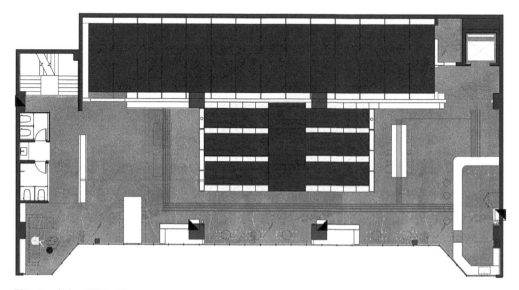

图5-2 书店一层平面图
Figure 5-2 First Floor Plan of Bookstore

5.2 Connection Mode Between Modules and Space at Different Scales

The connection mode between modules and space at different scales is mainly suitable for the spatial remodeling of interior space. The spatial remolding mentioned here refers to the remolding of the existing space by using the controllable boundary of the module to achieve the spatial composition that meets the design requirements. This connection method is suitable for improving space utilization, dividing functional areas and shaping specific space modeling. The commercial bookstore project and the Yantai Daily office space design project both use the connection mode between modules and space at different scales.

5.2.1 Commercial Bookstore Design Project

As can be seen from the floor plans, a black rectangular module was added to the north of the first floor in the commercial bookstore project, which reshaped the whole floor space and expanded the interior space of the first floor to two floors. This mode of spatial remolding according to the actual situation of interior space is a practical application case of the connection mode between modules and space of different scales, showing the role of this connection mode in improving space utilization. (Figure 5–2, Figure 5–3)

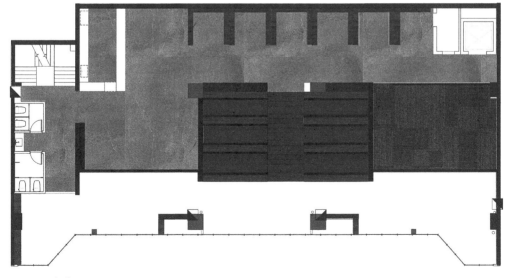

图5-3 书店二层平面图
Figure 5–3 Second Floor Plan of Bookstore

5.2.2 日报社办公空间设计项目

通过图5-4可以了解到，文化模块在整个办公空间中起到了划分功能区的作用。用模块的墙体边界代替了传统的墙壁对空间进行划分。另外两个模块和文化模块一起对主办公区进行了围合，根据空间尺度需要，对艺术模块和生态模块的体积进行了适度调整，发挥了模块与空间的不同尺度连接方式的划分功能区和塑造特定空间造型的作用。

■ 艺术模块　　■ 生态模块　　■ 文化模块

图5-4　日报社项目三个置入盒子视野分析图

5.2.2 Daily Office Space Design Project

It can be seen from the floor plan that the cultural module plays a role in dividing the functional areas in the whole office space. Modular wall boundaries are used instead of the traditional walls to divide the space. The other two modules, together with the cultural module, enclose the main office area. According to the spatial scale, the volume of the art module and the ecological module was adjusted appropriately, which gave play to the function of dividing the functional area and shaping the specific spatial modeling by the connection mode between modules and space of different scales.

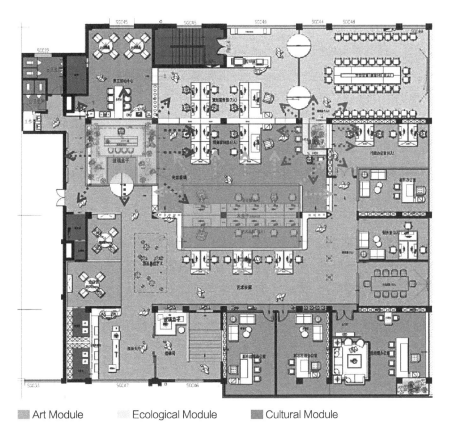

▨ Art Module   ▨ Ecological Module   ■ Cultural Module

Figure 5 – 4 Visual Field Analysis Diagram of Three Boxes

5.3 交通流线与视点连接方式

交通流线与视点的连接方式即本书开篇所提出的Spot – Belt模式，也是模块化建筑设计法的在连接方式上的创新点。针对商业书店项目这种文化融合受限的情况，模块化建筑空间的文化元素可以选用交通流线与视点连接的方式，即在每个模块中设立1～2个视觉中心点，利用交通流线把每个模块中的视点连接起来。这样可以在减少文化元素在空间当中占比同时，不影响整体文化氛围的营造。这种连接方式具有很强的匹配适应性，可以满足大部分的模块空间与文化元素融合的设计需求。为了让视觉中心点能够更好地吸引参观者的注意力，人的视觉特征分析成为交通流线与视点连接方式的理论基础。

5.3.1 视觉感知分析

视觉感知是人与空间之间联系的主要媒介。研究表明人类83%的信息来自视觉，11%来自听觉，其余6%来自嗅觉、味觉和触觉。因此，视觉是人们感知环境的最重要信息来源，它刺激人体产生了各种心理和行为反应。

视觉感知可以触发一系列心理活动。首先，通过视觉体验来体验建筑的空间尺度、形状、颜色和纹理。其次，视觉随着身体的运动而变化，并通过变化的视觉体验逐渐形成一系列连续的视觉图像。在设计过程中，有必要综合考虑视角、观察点、观察路线和观察距离。由于距离的差异，视觉体验会形成不同的视觉效果。最后，视觉本身有主动选择权，例如色彩强烈或体积巨大的事物会优先吸引视觉注意。

5.3 Connection Mode of Traffic Streamline and Viewpoint

The connection mode of traffic flow and viewpoint is the SPOT-BELT mode proposed at the beginning of this research, which is also the innovative point of modular architectural design method in the connection mode. In view of the limited cultural integration of the commercial bookstore project, the cultural elements of modular architectural space can choose the way of connecting traffic flow and viewpoint. In other words, 1-2 visual center points are set up in each module, and viewpoints in each module are connected by traffic flow lines. In this way, the proportion of cultural elements in the space can be reduced without affecting the overall cultural atmosphere. This connection mode has strong matching adaptability and can meet most of the design requirements of the integration of module space and cultural elements. In order to make the visual center better attract the attention of visitors, the analysis of human visual characteristics has become the theoretical basis of the connection mode between traffic flow and viewpoint.

5.3.1 Visual Perception Analysis

Visual perception is the main connecting medium between human and space. Studies show that people get their information 83 percent from sight, 11 percent from hearing, and the remaining 6 percent from smell, taste and touch. Therefore, vision is the most important source of information for people to perceive the environment, and it produces a variety of psychological and behavioral responses.

Visual perception can trigger a range of psychological activities. First, people can experience the spatial scale, shape, color and texture of the building through visual experience. Second, vision changes with the movement of the body and gradually forms a series of continuous visual images through the changing visual experience. In the design process, it is necessary to comprehensively consider the perspective, observation point, observation route and observation distance. Due to the difference in distance, the visual experience will form different visual effects. Third, vision has its active preference, such as strong colors or huge things which would attract visual attention first.

5.3.2 视野分析

人的视野在水平方向120°、垂直方向130°的范围内（即向上60°和向下70°）。所以，视点造型的设计要尽量保证在参观者120°的视野范围内。这样才能最大限度地增加视点造型的影响。

5.3.3 视点分析

视点也即观看者的位置。文化建模的观点应以多方向、多角度为特征。为了满足从各个角度吸引注意力的需求，应该从欣赏的角度和距离考虑。

5.3.4 方案解析

（1）商业书店设计项目

如图5–5所示，商业书店的一层地面设计有黄色金属质感的线条对顾客进行主要通道的路线指引。同时，在功能区的分流处，会有路面指示箭头进行标识。线条设计的目的，是要在顾客的视野可视范围内进行路线指引的心理暗示，引导顾客按照设计师的构想来游览整个室内空间，让空间中设立的视觉中心点能够发挥最大作用。（图5–6、图5–7）

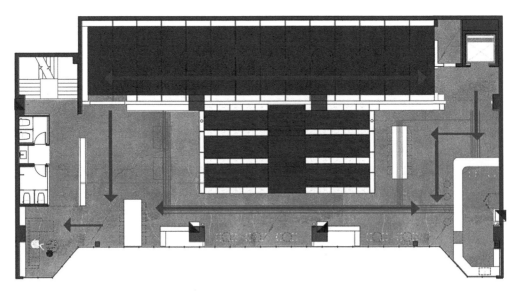

图5–5　书店一层交通流线图
Figure 5–5　First Floor Traffic Flow of the Bookstore

5.3.2 Visual Field Analysis

The human visual field is in the range of 120°
horizontally and 130° vertically, i.e. 60°up and
70°down. Therefore, the design of the viewpoint
shape should try to ensure that it can appear
within the range of 120° of the visitors. Only in
this way it can maximize the influence of the
view point modeling.

5.3.3 Viewpoint Analysis

Viewpoint refers to the position of the viewer.
The viewpoint of cultural modeling should be
characterized by multi-direction and multi-
angle. To satisfy the need of attracting attention
from all angles, the thinking should be taken in
terms of appreciation and distance.

5.3.4 Design Project Analysis

(1) Commercial Bookstore Design Project

The first floor of the commercial bookstore
is designed with yellow metal lines to guide
customers through the main channels. At the
same time, at the diversion of the functional
area, there will be a road indicator arrow
mark. The purpose of line design is act like
a psychological hint to guide the route within
the visual field of the customer, and guide
the customer to visit the whole interior space
according to the designer's idea, so that the
visual center point set up in the space can have
the maximum effect. (Figure 5-5~Figure 5-7)

图5-6 入口金属线指示效果图
Figure 5-6 The Rendering of the Entry Wire Indicates

　　设计师将书简造型作为视觉中心点布置于整个空间的中心区域，分别悬挂于阶梯书架的对面两侧。通过平面图可知，书简造型以整个空间的中心为轴线，平均布置在空间的东西两侧，这样可以最大限度地保证视觉中心点至少有一个被视野捕捉，从而激发空间的中国传统文化元素氛围感，最终达到以点带面的效果。这就是本书提出的应用策略的创新点，即Spot-Belt模式。

图5-7 分流处金属线指示效果图
Figure 5-7 The Rendering of the Metal Line Indicates at the Bypass Point

Designers put the bamboo slip shape in the central area of the whole space, respectively hanging on the opposite sides of the ladder bookshelves, as the two visual center points of the commercial bookstore space. It can be seen from the plan that the bamboo slip shape takes the center of the whole space as the axis and is evenly placed on the east and west sides of the space. In this way, at least one of the visual center points can be captured by the visual field to the maximum extent, so as to improve the atmosphere of Chinese traditional cultural elements in the space and finally achieve the effect of using a small visual point to drive the whole. This is the innovation point of this strategy, i.e. the spot-belt mode.

（2）烟台日报社办公空间设计项目

烟台日报社办公空间设计项目使用的连接方式与商业书店项目相同，同样是以金属指示线来引导来访人员和内部员工的动线，通过增加空间中的视觉中心点来烘托企业文化氛围。而与书店项目不同的是，日报社的项目设计将空间置入的模块自身作为了视觉中心点。也就是说，这个空间中的模块既承担了划分功能区的作用，又承担了视觉中心点的吸引作用。（图5-8、图5-9）

图5-8 办公区金属指示线效果图
Figure 5-8 Office Area Metal Indicator Line

(2) Yantai Daily Office Space Design Project

The Yantai Daily office space design project uses the same connection mode as the commercial bookstore project, which is the metal indicator line that guide the walking route of visitors and internal staff, and therefore enhances the corporate culture atmosphere by increasing the visual center point in the space. In contrast to the bookstore project, the Yantai Daily office design project places the space into the module itself as a visual center. That is to say, the modules in this space not only assume the function of dividing the functional area, but also assume the attraction function of the visual center point. (Figure 5–8, Figure 5–9)

图5-9 洽谈区金属指示线效果图
Figure 5-9 Negotiation Area Metal Indicator Line

总结

　　在中国乡村改造的需求和政府政策支持的背景下，新型模块化建筑在中国正在迅速发展。鉴于传统模块化建筑与设计难以协作，新的模块化建筑设计方法亟需理论支持和实践经验以及技术的不断创新。本书所提出的模块化设计法，相较传统模块化建筑而言，是一种更符合当下中国国情的设计方法。除了能够满足设计师的不同设计需求之外，还能够更好地与本土文化相融合。

　　笔者主要在模块化建筑的创新应用、文化承载方式、模块连接方式和影响本土文化元素占比因素等角度，对模块化建筑设计法进行了基于设计实践项目的理论探讨，希望能够为当代中国建筑领域增加一种能够将本土文化与当代建筑设计结合的设计方法。其中的创新点也希望能为其他的建筑设计方法提供新思路。同时，模块化建筑设计方法不仅仅适用于中国，对于其他一些已经掌握新型模块化建筑技术的国家来说，也具有参考意义。

　　然而，本书的设计案例数量不足，多为理论分析和主观描述，缺乏足量的研究样本和定量分析。在未来，对模块化建筑设计项目的评价研究会逐步增加，通过数据的量化会使研究结果更加准确，继而与本次的研究结果相互比较，进行验证和纠正。

Conclusion

New modular buildings are developing rapidly in China under the background of the demand for rural renovation and government policy support. Since it is difficult for traditional modular building and design to cooperate, new modular building design methods need to be supported by theory and innovated by practical experience. Compared with the traditional modular building, the modular design method proposed in this research is a design method more in line with the current situation of China. In addition to meeting the different design needs of designers, it can also better integrate with the local culture.

The author mainly discussed about the innovation of the modular construction, cultural load mode, module connection mode and influencing factors of the proportion of local cultural elements, and the modular architecture design method based on the theory of design practice project, hope to be able to increase a new design method that can combine the native culture and contemporary architecture design method. The innovation points in it also can inspire new ideas and other architectural design methods ideas. At the same time, modular building design method is not only applicable in China, for some other countries that have mastered the new modular building technology, this research could also be a significance reference.

However, the number of design cases in this research is insufficient, most of which are theoretical analysis and subjective descriptions, and lack of sufficient research samples and quantitative analysis. In the future, evaluation studies on modular architectural design projects will gradually increase. The quantification of data will make the research results more accurate, and then everyone could compare them with this research results for vevification and correction.

参考文献 Reference

[1] 刘颐佳，高路. 盒子结构建筑及应用与展望[J]. 四川建筑，2008，（05）：136-138.

[2] 陈莉. 装配式混凝土建筑建安工程造价研究[D]. 杭州：浙江大学，2018.

[3] 王从越. 基于BIM的装配式建筑模块化设计策略研究[D]. 重庆：重庆大学，2019.

[4] 李泳辰. 装配式背景下模块化住宅设计探究[D]. 青岛：青岛理工大学，2017.

[5] 刘博，张廷瑜，张挺，田兴. 住宅工业化设计及农村住宅建构研究[J]. 四川建材，2020，46（07）：29-30.

[6] 钟敦字. 产业化促进与普通住宅设计策略[D]. 重庆：重庆大学，2007.

[7] 郑彬，张玲玲. 烟台大学翼之队北方印宅[J]. 建设科技，2018，（15）：56-59.

[8] 任钟颖. 模块化设计在装配式住宅中的应用研究[J]. 重庆建筑，2021，20（S1）：69-72.

[9] 李燕红. 装配式建筑发展进程下的建筑设计策略[J]. 低碳世界，2021，11（07）：128-129.

[10] 时颖，康春霞. 绿色装配式钢结构建筑可持续发展路径分析[J]. 中小企业管理与科技（中旬刊），2021，（04）：126-127.

[11] 侯斯婕，陈蓉，王震. BIM技术背景下绿色建筑与装配式建筑融合发展的趋势研究[J]. 散装水泥，2021，（03）：62-64.

[12] 马祥. BIM技术背景下绿色建筑与装配式建筑融合发展的趋势研究[D]. 青岛：青岛理工大学，2018（45）：133-255.

[13] 刘奕麟. 模块化的设计思维在空间设计领域的研究与运用[J]. 设计，2021，34（14）：142-145.

[14] 李惠玲，王婷. 我国装配式钢结构住宅产业化发展面临的问题与对策研究[J]. 建筑经济，2020，41（03）：20-23.

[15] 胡泊，刘冰，韦凯杰. 钢结构建筑在装配式建筑发展过程中的优势[J].

福建建材，2020，（02）：31−32+106.

[16] 文强. 论述预制装配式建筑施工技术的研究与应用[J]. 居舍，2019，（36）：30.

[17] 沈克宁. 空间感知中的时间与记忆[J]. 建筑师，2015，（04）：48−55.

[18] 彭一刚. 建筑空间组合论[M]. 北京：中国建筑工业出版社，2008.

[19] 窦德嘏. 建筑空间体验中视线与路线的基本分析[J]. 泰州职业技术学院学报，2004，（03）：18-21.

[20] 王文静，李志武，于春义，刘洋，叶浩文，李振宝. 模块化钢结构建筑结构体系研究进展[J]. 施工技术，2020，49（11）：24−30+36.

[21] Bonaiuto, M., Bilotta, E., & Stolfa, A. Feng Shui and environmental psychology: A critical comparison[J]. Journal of Architectural and Planning Research, 2010, 27(1), 23−34.

[22] Combs, S. C. The Dao of rhetoric[M]. Albany: State University of New York Press, 2005.

[23] Gu, Z., & Lao, Z. The book of Tao and tech[M]. Beijing: China translation and Publishing House Press, 2006, 1−253.

[24] Knapp, R. G.China's living houses: Folk beliefs, symbols, and household ornamentation[M]. Honolulu: University of Hawaii Press, 1999.

[25] Morton, W. S., & Lewis, C. M. China: Its history and culture[M]. New York: McGraw−Hill, 2005.

图片来源

图 1−1 来源：https://doc.wendoc.com/ba9590210cb2bb9225542e0b7.html

图 1−2 来源：https://www.sohu.com/a/155611087_305341

图 1−3 来源：http://travel.qunar.com/p-pl5221918

图 1−4 来源：https://www.sohu.com/a/155504016_243901

图 1−5 来源：https://www.sohu.com/a/155504016_243901

图 1−6 来源：https://www.sohu.com/a/155504016_243901

图 1−7 来源：https://www.sohu.com/a/155504016_243901

图 1−8 来源：https://www.sohu.com/a/239380966_822566

图 1−9 来源：https://www.wendangwang.com/doc/32bf3c72abc6ea9b9681e509/6

图 1−10 来源：https://www.sohu.com/a/76063594_243901

图 1−11 来源：https://www.sohu.com/a/76063594_243901

图 1−12 来源：https://zhuanlan.zhihu.com/p/34910974

图 1−13 来源：https://zhuanlan.zhihu.com/p/34910974

图 1−14 来源：https://zhuanlan.zhihu.com/p/34910974

图 1−15 来源：Megan Sveiven 摄

图 1−16 来源：https://max.book118.com/html/2018/0719/8133136123001115.shtm

图 1−17 来源：http://www.360doc.com/content/15/1209/14/11881236_519094907.shtml

图 1−18 来源：http://www.360doc.com/content/15/1209/14/11881236_519094907.shtml

图 1−19 来源：http://huodong.fengniao.com/apply-4192.html

图 1−20 来源：http://www.sfjizhuangxiang.com/supply_details_1008194106.html

图 2−2 来源：郑彬提供

图 2−3 来源：郑彬提供

图 2−4 来源：郑彬提供

图 2−5 来源：郑彬提供

图 2−6 来源：郑彬提供

图 2−7 来源：郑彬提供

图 2−8 来源：郑彬提供

图 3−1 来源：郑彬提供

Source of picture

Figure 1–1 source: https://doc.wendoc.com/ba9590210cb2bb9225542e0b7.html

Figure 1–2 source: https://www.sohu.com/a/155611087_305341

Figure 1–3 source: http://travel.qunar.com/p-pl5221918

Figure 1–4 source: https://www.sohu.com/a/155504016_243901

Figure 1–5 source: https://www.sohu.com/a/155504016_243901

Figure 1–6 source: https://www.sohu.com/a/155504016_243901

Figure 1–7 source: https://www.sohu.com/a/155504016_243901

Figure 1–8 source: https://www.sohu.com/a/239380966_822566

Figure 1–9 source: https://www.wendangwang.com/doc/32bf3c72abc6ea9b9681e509/6

Figure 1–10 source: https://www.sohu.com/a/76063594_243901

Figure 1–11 source: https://www.sohu.com/a/76063594_243901

Figure 1–12 source: https://zhuanlan.zhihu.com/p/34910974

Figure 1–13 source: https://zhuanlan.zhihu.com/p/34910974

Figure 1–14 source: https://zhuanlan.zhihu.com/p/34910974

Figure 1–15 source: Photo by Megan Sveiven

Figure 1–16 source: https://max.book118.com/html/2018/0719/8133136123001115.shtm

Figure 1–17 source: http://www.360doc.com/content/15/1209/14/11881236_519094907.shtml

Figure 1–18 source: http://www.360doc.com/content/15/1209/14/11881236_519094907.shtml

Figure 1–19 source: http://huodong.fengniao.com/apply-4192.html

Figure 1–20 source: http://www.sfjizhuangxiang.com/supply_details_1008194106.html

Figure 2–2 Source: Zheng Bin offered

Figure 2–3 Source: Zheng Bin offered

Figure 2–4 Source: Zheng Bin offered

Figure 2–5 Source: Zheng Bin offered

Figure 2–6 Source: Zheng Bin offered

Figure 2–7 Source: Zheng Bin offered

Figure 2–8 Source: Zheng Bin offered

Figure 3–1 Source: Zheng Bin offered

图 3-2 来源：杨金生提供

图 3-3 来源：杨金生提供

图 3-4 来源：王润吉提供

图 3-5 来源：王润吉提供

图 3-8 来源：郑彬提供

图 3-9 来源：郑彬提供

图 3-10 来源：杨金生提供

图 3-11 来源：杨金生提供

图 3-12 来源：杨金生提供

图 3-13 来源：王润吉提供

图 3-14 来源：王润吉提供

图 3-15 来源：https://zhuanlan.zhihu.com/p/28635491?from_voters_page=true

图 3-16 来源：王润吉提供

图 3-17 来源：王润吉提供

图 4-1 来源：郑彬提供

图 4-2 来源：http://www.zhijian123.cn/m/view.php?aid=211

图 4-3 来源：郑彬提供

图 4-4 来源：杨金生提供

图 4-5 来源：杨金生提供

图 4-6 来源：杨金生提供

图 4-7 来源：杨金生提供

图 4-8 来源：杨金生提供

图 4-9 来源：王润吉提供

图 4-11 来源：王润吉提供

图 4-12 来源：王润吉提供

图 5-1 来源：郑彬提供

图 5-2 来源：王润吉提供

图 5-3 来源：王润吉提供

图 5-5 来源：王润吉提供

图 5-6 来源：王润吉提供

图 5-7 来源：王润吉提供

Figure 3-2 Source: Yang Jinsheng offered

Figure 3-3 Source: Yang Jinsheng offered

Figure 3-4 Source: Wang Runji offered

Figure 3-5 Source: Wang Runji offered

Figure 3-8 source: Zheng Bin offered

Figure 3-9 Source: Zheng Bin offered

Figure 3-10 source: Yang Jinsheng offered

Figure 3-11 Source: Yang Jinsheng offered

Figure 3-12 source: Yang Jinsheng offered

Figure 3-13 source: Wang Runji offered

Figure 3-14 source: Wang Runji offered

Figure 3-15 source: https://zhuanlan.zhihu.com/p/28635491?from_voters_page=true

Figure 3-16 Source: Wang Runji offered

Figure 3-17 source: Wang Runji offered

Figure 4-1 Source: Zheng Bin offered

Figure 4-2 source: http://www.zhijian123.cn/m/view.php?aid=211

Figure 4-3 Source: Zheng Bin offered

Figure 4-4 Source: Yang Jinsheng offered

Figure 4-5 Source: Yang Jinsheng offered

Figure 4-6 Source: Yang Jinsheng offered

Figure 4-7 Source: Yang Jinsheng offered

Figure 4-8 Source: Yang Jinsheng offered

Figure 4-9 Source: Wang Runji offered

Figure 4-11 source: Wang Runji offered

Figure 4-12 Source: Wang Runji offered

Figure 5-1 Source: Zheng Bin offered

Figure 5-2 Source: Wang Runji offered

Figure 5-3 Source: Wang Runji offered

Figure 5-5 Source: Wang Runji offered

Figure 5-6 Source: Wang Runji offered

Figure 5-7 Source: Wang Runji offered

后记

《模块化建筑设计的本土化应用策略》这本书的主要内容，提炼自我在匈牙利佩奇大学读博期间完成的博士论文。在此，本人重点感谢中国建筑工业出版社对我本人的信任和鼓励，感谢中国建筑工业出版社编辑为我提供的指点与建议；感谢所有为我的博士研究提供人力或专业技术支持的朋友、伙伴，并感谢马塞尔·布鲁尔博士学院（Marcell Breuer Doctoral School）提供让我继续自己博士研究的机会。

我特别要感谢我的导师Zoltán Erzsébet Szeréna教授和Akós Hutter教授，他们在我的论文准备阶段提供了许多有建设性的建议和指导。

我同样要感谢烟台大学郑彬老师和设计师杨金生、王润吉，感谢他们同意在我的著作中使用他们的设计项目，并提供资料和数据支持。

虽然在撰写本书的三年过程中尽己所能地去考察、调研、搜集和编写，由于知识与阅历有限，在行业的不断发展下，本书难免会有遗漏和不足之处。望各位专家学者、同仁、前辈多多包涵，同时也欢迎各位提供宝贵建议。

特别说明，本书中的部分图片及信息来自互联网，版权归原作者所有，本书仅供学习之用。在此，我也非常感谢这些无偿提供资料的平台与网友。本想找到资料的原作者表示感谢并征求意见，但由于有些资料经过多次转载，难以与原作者取得联系。因此，本人只有再次深深地感谢网络及所有帮助过我的网络朋友们。借本书做进一步学习推广的交流。

最后感谢我的家人在我博士学习期间和论文编写阶段的默默付出。

感谢所有！

Postscript

The main content of this book is extracted from my doctoral dissertation completed during my doctoral studies at The University of Pecs in Hungary. Here, I would like to thank China Building and Architecture Press for their trust and encouragement, and the editors of China Building and Architecture Press for their guidance and suggestions. I would like to thank all my friends and associates who provided support for my doctoral research, and the Marcell Breuer School for offering me the opportunity to continue my doctoral research.

In particular, I would like to thank my supervisors, Professor Zoltán Erzsébet Szeréna and Professor Akós Hutter, who provided many constructive suggestions and guidance during the preparation of my thesis.

I would also like to thank Professor Zheng Bin from Yantai University, designer Yang Jinsheng and designer Wang Runji for agreeing to use their design projects in my work and providing materials and data support.

Although I have tried my best to investigate, investigate, collect and compile this book during the three years of writing, due to my limited knowledge and experience, there will inevitably be omissions and deficiencies in this book under the continuous development of the industry. I look forward to the experts and scholars, colleagues and predecessors, as well as your valuable suggestions.

Special note, some of the pictures and information in this book come from the Internet, the copyright belongs to the original author, this book is only for learning use. Here, I am also very grateful to these platforms and netizens who are free to provide information. I would have liked to find the original author of the information to express gratitude well and for comments, but because some information after many reprints, it is difficult to get in touch with the original author. Therefore, I can only deeply thank the network and all the network friends who helped me. Through this book for further study and promotion of communication.

Finally, I would like to thank my family for their silent efforts during my doctoral study and thesis writing.

Thank you all!

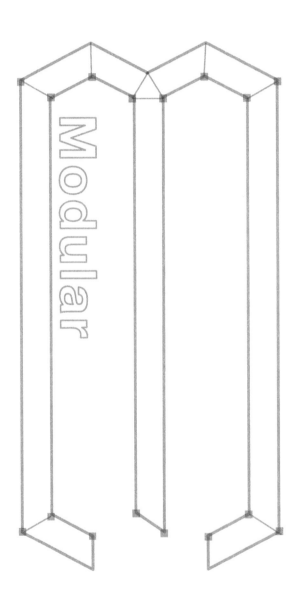